MATHEMATICS RESEARCH DEVELOPMENTS

CONTROL THEORY AND ITS APPLICATIONS

MATHEMATICS RESEARCH
DEVELOPMENTS

Additional books in this series can be found on Nova's website
under the Series tab.

Additional E-books in this series can be found on Nova's website
under the E-book tab.

MATHEMATICS RESEARCH DEVELOPMENTS

CONTROL THEORY AND ITS APPLICATIONS

VITO G. MASSARI
EDITOR

Nova Science Publishers, Inc.
New York

LIBRARY OF CONGRESS CATALOGING-IN-PUBLICATION DATA

Control theory and its applications / editor, Vito G. Massari.
     p. cm.
 Includes bibliographical references and index.
 ISBN 978-1-61668-384-9 (hardcover)
 1. Control theory. 2. Automatic control. 3. Self-control. 4. Human behavior. 5. Electric motors--
Automatic control. I. Massari, Vito G.
 QA402.3.C6395 2010
 629.8'312--dc22
                              2010047109

Published by Nova Science Publishers, Inc. ✦ New York

CONTENTS

PREFACE

Control theory is a theory that deals with influencing the behavior of dynamical systems and an interdisciplinary subfield of science, which originated in engineering and mathematics, and evolved into use by the social sciences, such as psychology, sociology and criminology. In this book, the authors present and discuss topical data on control theory relating to these fields.

As explained in Chapter 1, self-regulation involves overriding or inhibiting behaviors, urges, or desires that would otherwise interfere with goal directed behavior (Barkley, 1997; Baumeister, Heatherton & Tice, 1994). As such, the ability to exert self-control is fundamental to successful smoking cessation. An innovative model of self-control, the *Self-Control Strength Model* (Muraven & Baumeister, 2000) offers a theoretical explanation for potential failures of self-regulation, suggesting that when a limited resource for volitional activities is depleted by exerting self-control on one task, people are less effective in subsequent self-regulation efforts. Study 1 demonstrates that participants whose self-control strength is depleted by resisting eating from a plate of tempting sweets are more likely to smoke immediately afterwards. Study 2 examines an intervention designed to offset these detrimental effects and replenish depleted self-control strength by inducing positive affect. Participants whose self-control strength was depleted by resisting sweets, but were then induced with positive emotions were able to counteract the effects of depletion and were less likely to smoke immediately afterwards. These findings may be particularly important for individuals who desire to diet while attempting to quit smoking out of concern for weight gain. Implications for tobacco cessation interventions are discussed.

As discussed in Chapter 2, problems relating to self-regulatory skills, interpersonal skills, and learning difficulties place children at increased risk for persistent externalizing behaviors. However, less is known about the antecedents of externalizing behavior in children most at-risk for antisocial behavior over the life-course. The current study used longitudinal data from 1,594 children previously shown to have severe behavioral problems selected from the Early Childhood Longitudinal Survey, Kindergarten Class (ECLS-K) to examine these developmental pathways. Structural equation modeling showed that learning problems, fine motor problems, and gross motor problems occurring at wave 1 were interrelated and variously predictive of self-control deficits and interpersonal deficits at wave 2. Both self-control and interpersonal deficits at wave 2 significantly predicted externalizing behaviors at wave 4. The findings add to an accumulating knowledge base indicating that externalizing behaviors are importantly related to learning, motor, interpersonal, and self-control deficits.

For dynamical systems described by parabolic differential equations, a problem of the tracking of a trajectory (a problem of etalon motion tracking) and a problem of the tracking of a control (a problem of dynamical input identification) are investigated in Chapter 3. Solving algorithms stable with respect to informational noises and computational errors are designed. The algorithms are based on the method of feedback control.

Chapter 4 deals with the parametric nonlinear suboptimal control of divergent hyperbolic equations with pointwise state constraints. Based on the ideology of the perturbation method, the authors study the suboptimal control problem depending on the infinite–dimensional parameter which is additively contained in the pointwise inequality constraints.

The minimizing sequence of usual admissible controls is the basic element of the authors' theory. Accordingly, as a minimizing sequence, the authors use the so-called minimizing approximate solution in the sense of Warga, but not classical minimizing sequence. The authors study the following issues: finding of suboptimality, regularity, and normality conditions; perturbation theory problems; differential properties of value functions (S–functions); optimal values stability (sensitivity). Detailed examination of these issues as applied to optimal control problems with pointwise state constraints is important both from the standpoint of the general theory of distributed optimization and from the standpoint of constructing various efficient numerical algorithms for solving optimization problems for such systems. For calculation of functionals' first variations, the authors use the so-called two– parametrical needle variation of admissible controls. Besides, the authors discuss an approximation of the source problem by problems with a finite number of functional constraints.

In Chapter 5, a novel sensorless adaptive robust control law is proposed to improve the trajectory tracking performance of induction motors. The proposed design employs the so called vector (or field oriented) control theory for the induction motor drives and the designed control law is based on an integral sliding-mode algorithm that overcomes the system uncertainties. The proposed sliding-mode control law incorporates an adaptive switching gain to avoid calculating an upper limit of the system uncertainties. The proposed design also includes a new method in order to estimate the rotor speed. In this method, the rotor speed estimation error is presented as a first order simple function based on the difference between the real stator currents and the estimated stator currents.

The stability analysis of the proposed controller under parameter uncertainties and load disturbances is provided using the Lyapunov stability theory. Finally simulated results show, on the one hand that the proposed controller with the proposed rotor speed estimator provides highperformance dynamic characteristics, and on the other hand that this scheme is robust with respect to plant parameter variations and external load disturbances.

In: Control Theory and Its Applications
Editor: Vito G. Massari, pp. 1-40

ISBN: 978-1-61668-384-9
© 2011 Nova Science Publishers, Inc.

Chapter 1

SELF-CONTROL STRENGTH THEORY AND ITS APPLICATIONS TO SMOKING BEHAVIOR

Dikla Shmueli

University of California, San Francisco

Abstract

Self-regulation involves overriding or inhibiting behaviors, urges, or desires that would otherwise interfere with goal directed behavior (Barkley, 1997; Baumeister, Heatherton & Tice, 1994). As such, the ability to exert self-control is fundamental to successful smoking cessation. An innovative model of self-control, the *Self-Control Strength Model* (Muraven & Baumeister, 2000) offers a theoretical explanation for potential failures of self-regulation, suggesting that when a limited resource for volitional activities is depleted by exerting self-control on one task, people are less effective in subsequent self-regulation efforts. Study 1 demonstrates that participants whose self-control strength is depleted by resisting eating from a plate of tempting sweets are more likely to smoke immediately afterwards. Study 2 examines an intervention designed to offset these detrimental effects and replenish depleted self-control strength by inducing positive affect. Participants whose self-control strength was depleted by resisting sweets, but were then induced with positive emotions were able to counteract the effects of depletion and were less likely to smoke immediately afterwards. These findings may be particularly important for individuals who desire to diet while attempting to quit smoking out of concern for weight gain. Implications for tobacco cessation interventions are discussed.

Introduction

A physician implores his patient to make some vital lifestyle changes after suffering a heart attack, in order to prevent another heart attack, heart disease or stroke. His advice for preventing these potential medical problems includes quitting smoking, becoming more physically active, and resisting the temptation for overindulging in high cholesterol foods and alcoholic beverages. Similarly, as New Year's Eve approaches an optimistic couple vows to quit smoking, begin a diet, and be more patient with their respective in-laws. These are just some examples of situations in which an individual desires to simultaneously change several behaviors that require self-control.

As anyone who has tried to quit smoking, go on a diet, or struggle to get to a gym on a cold evening after a long day at work, will intuitively know: it is no small feat. Previous research has demonstrated that these attempts at inhibition and persistence are linked to trait or dispositional levels of self-control (Tangney, Baumeister & Boone, 2004). The current chapter extends this research by integrating an innovative theory of state or momentary self-control, the *self-control strength model* to explain the difficulties in engaging in more than one behavior change.

The self-control strength model suggests that individuals have a limited amount of resources available for tasks that require self-control such as inhibiting a behavior or delaying immediate gratification (Baumeister, Bratslavsky, Muraven & Tice, 1998; Muraven & Baumeister, 2000). Using self-control on one task (e.g., resisting a second beer) consumes or depletes this resource and leads to poorer self-control on a subsequent self-control task. Thus, inhibiting one behavior may impair the other. This may be particularly detrimental for smokers who are trying to quit and have additional self-control demands. Studies show that an overwhelming majority of tobacco users (92%) do in fact engage in a least one additional health risk behavior (Fine, Philogene, Gramling, Coups & Sinha, 2004; Klesges, Eck, Isbell, Fulliton & Hanson, 1990).

The current chapter will review a series of studies demonstrating that resisting the temptation of sweets may lead to greater likelihood of subsequent smoking. The second study will also present a positive affect intervention designed to replenish self-control strength and counteract the effect of resisting sweets on subsequent smoking. These findings have implications for smokers who desire to diet while they are simultaneously attempting to quit smoking. Broader implications for multiple health behavior change are discussed as well.

Self-control

Defining Self-control

The concept of self-control, also referred to as willpower, self-discipline, and impulse control in different contexts and fields, has been of interest throughout history. The ancient Greeks used the term "akrasia" to define the state of lacking command over oneself or acting against one's better judgment, and many are familiar with the quote by Aristotle "I count him braver who overcomes his desires *than him who conquers his enemies; for the hardest victory is over self.*"

Contemporary definitions of self-control include overriding or inhibiting initial impulses, thoughts, or desires that would otherwise interfere with goal directed behavior (Barkley, 1997; Baumeister, Heatherton & Tice, 1994). Individuals exert self-control to follow a rule (either internally or externally determined) or to delay immediate gratification (Hayes, Gifford & Ruckstuhl, 1996). Without self-control, an individual would carry out his or her normal, typical, or automatic response tendencies and engage in immediate, short-term focused actions (Rachlin, 2000). Throughout the chapter I will use the terms self-control and self-regulation interchangeably. However, those who make a distinction generally consider self-regulation a broader set of processes including both conscious deliberate efforts (e.g., resist eating a piece of cake) as well as unconscious homeostatic processes (e.g., regulating body temperature), whereas self-control refers solely to the former.

Self-regulation is an extremely important executive function of the brain and deficits in self-control or self-regulation are implicated in many psychological disorders such as ADHD (Barkley, 1997). Self-control is also a uniquely human trait, setting us apart from other animals. Some propose evolutionary explanations, suggesting that self-control was particularly important and adaptive for humans as they formed cultural groups (Baumeister, 2005). The capacity to self-regulate develops early in life. Research demonstrates that during the preschool period children progress in their self-control abilities (Kochanska, Murray & Harlan 1990). This is particularly important due to the increasing number of activities requiring self-control, such as waiting to play with a desired toy, or sitting quietly in class. The following sections will provide two important perspectives to viewing self-control that are useful in their unique and combined contributions and applications to both smoking and weight management.

Trait Conceptions of Self-control

Historically, self-control or willpower has been conceptualized as a stable disposition or trait. That is, individual differences in self-control have been viewed as reflections of a global stable disposition, part of a person's overall personality. An individual with a trait of high self-control (i.e., having willpower) is one who successfully regulates his or her impulses, follows societal rules, and resists temptations in different contexts and over time, whereas a person with low self-control (i.e., lacking willpower) will have a general tendency to fail at regulating these impulses. The conceptualization of self-control as a disposition or trait is intuitive and appealing. We all know people who are better able to hold their tempers, resist having seconds at meals or a dessert, and persevere at even the most tedious and frustrating tasks. The distinction between individuals based on their inherent stable self-control has been defined and measured in numerous ways, and examined as it relates to other person level variables and over time.

One perspective on trait self-control has been to conceptualize it in terms of the ability to delay gratification, that is, to postpone immediate gratification for the sake of delayed but more valued outcomes (Mischel, 1974). In a series of studies using a delay of gratification paradigm, individual differences in the ability to forgo immediate gratification were found as early as the preschool years. In what is often referred to as the "Marshmallow Experiment", preschool children were left by themselves in a room and were given a plate with a single marshmallow on it and were told that if they waited and didn't eat it until the experimenter returned they would get another marshmallow as a bonus. Thus, they were placed in a situation that provided a choice between receiving a smaller immediate reward or a larger delayed reward. Findings demonstrated that some children were able to wait and delay their and others could not. Their capacity to wait is reflective of their self-control, their ability to delay gratification. This study was later replicated numerous times with different variations but the same principle of choice between a small immediate reward and a delayed but more valuable reward (e.g., a greater quantity of candy). Researchers then followed the progress of the children into adolescents and found that being able to wait for a preferred treat was predictive of their later self-regulatory and coping skills as adolescents (Mischel, Shoda & Peake, 1988; Mischel, Shoda & Rodriguez, 1989; Shoda, Mischel & Peake, 1990). These longitudinal studies revealed that those who delayed gratification longer as children were rated by their parents as having better self-control as adolescents, more able to resist temptation, cope with frustration and stress, and even had higher verbal and quantitative SAT scores. Furthermore, in a later follow up when these individuals were in their early thirties, the links remained significant between their preschool delay of gratification and adult self-regulatory abilities (Ayduk et al., 2000). In that sense, although Mischel and his colleagues stress that the circumstances greatly influence how a trait is expressed in any particular situation (Mischel & Shoda, 1995), in the long run self-regulation seems to be relatively consistent over long periods of time.

Other research supports the link between trait self-control and behavior in numerous other aspects of life. In fact, failure to adequately regulate behavior lies at the core of many personal and social problems, such as debt and bankruptcy, unwanted pregnancy, AIDS and other sexually transmitted diseases, and substance abuse problems (see Baumeister, Heatherton & Tice, 1994 for review). For example, men with better self-control are shown to be less likely to get divorced (Kelly & Conley, 1987). Self-regulation failure has also been associated with the ability to diet or regulate food intake (Herman & Polivy, 1975).

Trait self-control has been measured in various ways. Behavioral measures are used in tests of delay of gratification discussed previously. There are also self-report scales designed to capture these stable individual differences in self-regulation. The Self-Control Questionnaire was (Brandon, Oescher & Loftin, 1990) is a dispositional self-control scale emphasizing self-control of health behaviors such as eating patterns. The Self-Control Schedule, developed by Rosenbaum (1980), is another self-control scale, however it is designed specifically for use with clinical samples. As such it emphasizes use of strategies such as self-distraction to solve various behavioral and clinical problems. The scale has good validity (e.g., Richards, 1985) but it not appropriate as a trait measure of self-control across broad spheres of normal behavior. Other researchers aim at assessing conscientiousness as one dimension of the Big Five aspects of personality (Costa & McCrae, 1992; Gosling, Rentfrow & Swann, 2003). The Rosenbaum Self-Control Scale (SCS, Rosenbaum, 1980) is a 36-item measure of self-control behaviors classified into four particular areas: (1) emotion regulation, (2) problem-solving, (3) ability to delay gratification, and (4) self-efficacy. Items are rated on a 6-point Likert scale ranging from (3) *very characteristic of me*, to (-3) *very uncharacteristic of me*. Sample items include: "When I do a boring job, I think about the less boring parts of the job and the reward I will receive when I finish", and "When I have something to do that is anxiety arousing for me, I try to visualize how I will overcome my anxieties while doing it". The scale has demonstrated good test-retest reliability and internal consistency based on six different samples (Rosenbaum & Jaffe, 1983)

A more recent 36-item trait scale of self-control developed by Tangney, Baumeister, and Boone (2004) allows for further measure of individual differences in self-control across normal samples and a wide range of behaviors. People who scored lower in self-control based on this self-control scale reported lower grade-point averages in college as well as more eating disorders and alcohol abuse problems (Tangney, Baumeister & Boone, 2004).

State Conceptions of Self-control

Another perspective or conceptualization of self-control involves viewing it as a state rather than trait level variable. That is, instead of viewing an individual's self-control as representing a stable personality trait or ability, this perspective focuses on temporary fluctuations in self-control. These two perspectives combined create a full picture that can help explain and account for behaviors which rely on self-control. People can be classified as generally high or low in trait self-control, but at any given moment people will have a certain capacity to exert self-control, based on temporary fluctuations around their average level of ability. The *Self-Control Strength* model introduces this phenomenon and offers an explanation for these fluctuations by suggesting that the level of self-control a person has at any given time is directly related to the amount of self-control he or she has recently exerted (Muraven, Tice & Baumeister, 1998).

Specifically, the self-control strength model proposes that people have an internal resource, self-control strength, which is necessary for the executive functioning of the self, which is the part of the self that makes decisions, resists temptations, overrides impulses, and generally controls the conscious and intentional behavior of the self. Thus, all acts of volition and self-control (e.g., going on a diet, persisting on a difficult or frustrating task) require and consume this same strength or resource. Another important characteristic of this resource is that it is finite or limited in nature. Similar to a strength or muscle it may become tired and its strength depleted after use (Muraven, Tice & Baumeister, 1998; Baumeister, Muraven & Tice, 2000; Baumeister, Bratslavsky, Muraven & Tice, 1998 Schmeichel, Vohs & Baumeister, 2003). For example, a person on a diet who is constantly regulating their nutritional intake and potentially resisting the desire to have various unhealthy or fattening foods throughout the day will gradually deplete this reserve of self-control strength and have less left for other self-control demands. Because the strength is limited, and becomes depleted with use, the self-control strength model predicts that exerting self-control on any given task will impair people's performance on a subsequent task of self-control. For instance, an alcoholic who has recently quit and resists the impulse to have a drink throughout the day is depleting his self-control strength and may be more likely to succumb to the temptation of a rich cake for dessert, despite being on a diet. Thus, two seemingly unrelated tasks that demand self-control can affect one another, because they both require and deplete the same strength or resource.

A direct measurement of self-control strength is not possible with the current level of technology. The conceptualization of self-control as like a muscle, or a limited resource or pool of energy is a useful metaphor designed to understand the mechanism underlying this important strength. The precise way of measuring this energy resource in yet unknown, however, psychologists have assessed the model indirectly by manipulating self-control using a variety of self-control depletion tasks and empirically testing the effect on subsequent self-control performance. These experiments have followed the classical depletion paradigm, which involves depleting self-control by having individuals exert self-control in one domain and then measuring their self-control in a different domain. For instance, in one study participants who were instructed to suppress their emotions while watching a gruesome film performed more poorly on a subsequent handgrip task of persistence (Muraven, Tice & Baumeister, 1998). Consistent with the self-control strength model, numerous studies using this paradigm have demonstrated that exerting self-control impairs further self-control performance in a wide variety of tasks and behaviors (e.g., Baumeister, Bratslavsky, Muraven & Tice, 1998; Vohs & Heatherton, 2000). These studies have used an assortment of self-control depletion techniques (i.e., initial tasks that require self-control) and have assessed subsequent self-control performance in a variety of areas ranging from persistence tasks to alcohol consumption. Thus, the effects of depleted self-control strength have been applied to many important areas such as addictions, stereotypes and prejudice, criminal behavior, and more recently, as demonstrated in later sections, smoking behavior.

Self-control Depletion and Interpersonal Relationships

Successful regulation of emotions and behaviors is essential for interpersonal contact between individuals and within groups. On any given day there are numerous interactions with others that require individuals to exert self-control. Some routine examples may include controlling one's temper in a fight with a spouse, restraining from retaliation at a co-worker, or putting on a happy face to celebrate a friend's birthday despite having had a bad day. Failure to exercise appropriate self-control can have detrimental consequences. The self-control strength model has been applied to a variety of behaviors that impact interpersonal relationships such as emotion regulation, self-presentation, stereotyping and prejudice, interpersonal conflict, sexual behavior, and aggression. In one study participants whose self-control was depleted by exerting self-control on an initial task were less able to control their emotional expressions (Muraven, Tice & Baumeister, 1998). Participants were randomly assigned either to the depletion condition in which they were asked to suppress thoughts of a white bear, or to the control condition, which did not require self-control exertion. Then they viewed a brief comedy video clip consisting of skits taken from different programs such as Saturday Night Live, and instructed to control their normal emotions and in particular avoid showing any signs of amusement while watching the video clip. Participants' facial expressions were videotaped and later coded. The results revealed that participants who were depleted were less able to regulate their emotional expressions. Specifically, they smiled more frequently and were rated as expressing more amusement overall while watching the film. Thus, depleted self-control resources impair the ability to successfully regulate emotions which is essential for harmonious interpersonal interactions.

Depleted self-control resources can also have deleterious consequences for stereotyping and prejudicial behavior. A recent study demonstrated that individuals who were more depleted were less able to inhibit automatic racial biases (Govorun & Payne, 2006). After exerting self-control in a preliminary task they completed a weapon identification task, which served as a measure of racial stereotyping. In this second task participants saw images of different weapons, which were preceded by brief presentations of Black or White faces, and were asked to quickly identify each object. Participants who were depleted and had a strong automatic bias tended to misidentify the harmless objects as weapons when they were followed by a Black face. These results may help explain situations in which biases may affect behavior, and have important social implications. For example, research has shown that a police officer's decision to shoot may be biased by racial factors (e.g., Greenwald, Oakes & Hoffman, 2003). Self-control depletion may be an additional factor that increases the likelihood of acting on automatic stereotypes and beliefs.

Other areas that have been explored using the self-control depletion paradigm include interpersonal conflict and aggressive behavior. In a series of experiments Stucke & Baumeister (2006) demonstrated that self-control depletion impaired participants' ability to inhibit aggressive behavior. In one study participants were depleted by resisting the urge to eat from a plate of tempting sweets and then their reaction to an insult was measured. This was done using a behavioral manipulation and outcome, whereby the experimenter in the study made an insulting remark to the participant, and later in the study the participants were given the opportunity to rate the experimenter who had insulted them. Those who had depleted their self-control strength earlier in the study were more likely to retaliate by giving a negative and potentially damaging evaluation of the experimenter, whereas those who were not depleted of self-control were more likely to inhibit this act of aggression or retaliation against the experimenter. These findings suggest that aggressive behavior can be restrained, but this restraint may be weakened when self-control has been depleted by a prior act.

A recent set of studies investigated the effect of self-control depletion on a variety of helping behaviors. Helping others requires self-control in order to override the selfish impulses that favor not helping others. Participants in the three studies were depleted of self-control using an assortment of manipulations (e.g., breaking a habit, controlling their attention while watching a video clip). Their pro-social or helping behavior was then assessed by self-report willingness to help in a hypothetical scenario, or volunteering to help a victim of a recent tragedy. Results of the studies demonstrated that participants who were depleted were less likely to help others. (Dewall et al., 2008).

Another series of studies investigated self-control in the context of sexual restraint and demonstrated that individuals who exert and deplete their self-control resources may be more likely to exhibit sexual behaviors that are normally restrained such as inappropriate sexual thoughts and the desire to engage in sexual activities with someone other than their primary relationship partner (Gailliot & Baumeister, 2007). In one study participants were randomly assigned either to a control group or to the experimental group that was instructed to complete a task of inhibition designed to deplete self-control. They were then given a series of word stems and anagrams to complete, both of which could be solved with either sexual or nonsexual words. For example, the word stem S _ _ could be solved as SEX or SET, and the anagram NISEP could be solved as PENIS or SPINE. Results demonstrated that depleted participants (as well as those with lower trait self-control) were more likely to respond with sexual words, even though social norms would suggest that people refrain from expressing sexual thoughts in these types of social situations. Another laboratory based study explored whether self-control depletion would influence participants' actual sexual behavior and restraint. Twenty one romantic couples participated in the study and were randomly assigned either to a task that involved inhibition (designed to deplete their self-control resources) or to a control task that did not require exerting self-control. They were then given 3 minutes of privacy in the lab and asked to express some sort of physical intimacy that they felt comfortable with (e.g., holding hands, hugging). Results showed that couples who were depleted were less likely to restrain their sexual behavior subsequently and engaged in more extensive behavior (e.g., groped each other and even removed articles of clothing). These results were found particularly among sexually inexperienced couples that had not yet established a comfortable level of sexual intimacy.

Self-control Depletion and Crime

Self-control depletion has also been shown to affect deviant and criminal behaviors. A series of three laboratory experiments investigated the influence of self-control strength on the likelihood of engaging in deviant behavior. In these laboratory experiments participants' self-control strength was depleted by having to engage in an initial task that required and consumed self-control strength. Some participants were then given an opportunity to commit a deviant act, which was to cheat on an additional task that required self-control. Cheating is a dishonest deviant act undertaken in pursuit of self-interest, which involves failure to inhibit or restrain the desire to follow a socially inappropriate impulse. Thus, although cheating may be a relatively minor act in the hierarchy of criminal behavior, it is an ideal representation of the fundamental self-control failures involved in crime.

In one experiment participants were depleted by exerting self-control on an initial thought suppression task. Previous research has demonstrated that suppressing thoughts is effortful (Wegner, Schneider, Carter & White, 1987) and requires self-control (Muraven, Collins, Nienhaus, 2002). Participants subsequent self-control performance was then assessed by having them complete a difficult cognitive anagram solving task. Unbeknownst to them, the anagrams were actually unsolvable. Participants were told that the experimenter did not need the anagram sheets back and that they could throw them away when the computer indicated that their time was up. They should simply indicate how many they solved on the computer. Because they were not required to disclose the "evidence" this design provided participants with an opportunity to cheat and exaggerate the number of anagrams they reported solving. The results of the study showed that participants who suppressed the thought of a white bear in the first part of the experiment were more likely to falsely report solving the (unsolvable) anagrams than those who were in the control condition (Muraven, Pogarsky & Shmueli, 2006). A second study replicated these results but used different measures of cheating. Specifically, after a self-control depletion task participants were given the opportunity to cheat by breaking two potential rule violations: reporting solving anagrams that were actually unsolvable (similar to study 1) or amount of time spent trying to solve the problem after the computer program prompted them that their time was up and that they must continue to the next screen. Participants whose self-control had been depleted in the initial task of self-control were more likely to work longer on the logic problems after being ordered to stop and also to lie regarding the number of problems they had solved as compared to participants who were not depleted.

Self-control Depletion and Addictive Behaviors

The self-control strength model is particularly useful for understanding addictive behaviors and has been applied to a variety of these behaviors including alcohol consumption and excessive eating. In one study, male social drinkers whose self-control strength was depleted in an initial task consumed more alcohol in a situation that demanded restraint (Muraven, Collins & Nienhaus, 2002). The study was designed so that participants thought it was about perceptions of intoxication, which were important because people who drink must judge whether they are capable of driving safely. The participants were told that they would sample two different beers and then take a driving simulator test, and that if they did well on the driving test they could win a prize. They were then randomly assigned to complete either a depleting thought suppression task or a control task that did not involve self-control exertion. The men in both conditions then entered a laboratory bar which was set up to look like a real bar. They were presented with a pitcher of Budweiser beer, a pitcher of Beck's beer and two empty glasses and were asked to rate the two beers on a variety of things such as sweetness. The situation called for them to restrain the amount of beer they tasted, because alcohol would impair their performance on the subsequent driving test they would take, as well as decrease their chances of winning a prize for good performance on the test. Despite these incentives to limit their drinking, the participants who had been depleted of self-control strength (by the earlier thought suppression task) consumed more beer during the taste test and had a higher blood alcohol content than those in the control group. The findings indicate that the amount of alcohol consumed in situations that require restraint is influenced in part by self-control strength as determined by previous self-control demands.

Another study investigated daily fluctuations in self-control demands in a sample of underage social drinkers, using ecological momentary assessment (EMA) (Muraven, Collins, Shiffman & Paty, 2005). Participants were given hand-held computers used for self-monitoring behavior to carry with them for about three weeks. They were instructed to complete a brief assessment each time they had a drinking episode, and were also prompted to answer various questions about their mood, location, current activities (e.g., working, driving, leisure) and other self-control demands (e.g., emotion regulation, dealing with stress) throughout the day. Each night they were also asked whether they planned to drink in the future and whether they planned to limit their drinking. Results demonstrated that on days that participants experienced more self-control demands, they were more likely to violate the drinking limit they set out for themselves. The results suggest that when participants exerted self-control on demands other than drinking (e.g., controlling their moods, thoughts or behaviors), it impaired their subsequent ability to control their drinking according to their self-imposed limits.

Another study which investigated self-control in the domain of alcohol use reversed the direction of influence and examined the effect of resisting alcohol on subsequent self-control performance in unrelated tasks (Muraven & Shmueli, 2006). Participants in the study were social drinkers who came in to the laboratory to take part in a controlled randomized study investigating how exposure to various substances affects performance. They were exposed to the sight and smell of water (the control condition) and their favorite alcoholic beverage (the experimental condition, designed to deplete their self-control strength) in different trials, but asked to resist drinking either beverage. Resisting the temptation to drink the alcohol required self-control and thus depleted their self-control strength. Their self-control strength was then measured on two subsequent tests of self-control, a handgrip task and a computerized task of inhibition. Because different acts that require self-control draw on the same pool or resource, these acts are interchangeable so that demands in one domain should lead to subsequent decrements in self-control in other areas. Results supported the self-control strength model and demonstrated that after resisting their favorite alcoholic beverage participants' performance on the subsequent tasks of self-control was weakened. This study establishes that depleting self-control by restraining drinking impairs subsequent self-regulatory performance on other tasks in the same way that performing tasks that require self-control impairs subsequent drinking restraint.

The two studies to be presented in this chapter apply the self-control strength model to another area of addictions- smoking behavior. The two studies were designed to demonstrate that exerting self-control on one task (dietary restraint) may impair subsequent attempts to control smoking. A brief section on tobacco use and weight management precede the discussion of the two studies, and provide a useful framework for understanding the framework and dynamics underlying the interplay between these two regulatory demands.

Tobacco Use

Smoking prevalence in the United States has been declining and in 2007 just under 20% of adults in the US were cigarette smokers (Centers for Disease

Control, 2007). This figure is slightly higher (23%) among high school students, and in middle school about 8% of children smoke. Furthermore, about 8% of high school students and 3% of middle school students in the United States are smokeless tobacco users (Centers for Disease Control and Prevention, 2006).

Table 1. Some of the carcinogens found in tobacco smoke

Chemical Name	Use or Characteristic
Arsenic	A heavy metal toxin
Benzyne	A chemical found in gasoline
Beryllium	A toxic metal
Cadmium	A metal used in batteries
Chromium	A metallic element
Ethylene Oxide	A chemical used to sterilize medical devices
Nickel	A metallic element
Polonium-210	A chemical element that gives off radiation
Vinyl Chloride	A toxic substance used in plastics manufacture

Tobacco smoke contains chemicals that are harmful to smokers and to nonsmokers who may be affected by secondhand smoke exposure. There are at least 250 dangerous chemicals in tobacco smoke, among them hydrogen cyanide (used in chemical weapons), carbon monoxide (found in car exhaust), formaldehyde (used as an embalming fluid), ammonia (used in household cleaners), and toluene (found in paint thinners). Of the 250 dangerous and harmful chemicals in tobacco smoke, over 50 have also been found to cause cancer. Table 1 provides information on some carcinogens (for more information see National Toxicology Program, 2005, National Cancer Institute, 1999).

Tobacco use is considered the leading preventable cause of death, accounting for about 5 million deaths a year worldwide (World Health Organization, 2006). In the US alone there are over 400,000 deaths per year. Of these about 40% are from cancer, 35% are from heart disease and stroke, and 25% from lung disease (Centers for Disease Control, 2005). Furthermore, secondhand smoke leads to about 38,000 deaths annually in the United States (Centers for Disease Control, 2008). Smoking is Tobacco is also one of the strongest cancer-causing agents and is linked with various different cancers and cardiovascular diseases. Lung cancer is the leading cause of cancer death in the United States, and is largely attributed to smoking: about 90% among men, and 80% among women (U.S. Department of Health and Human Services, 2004). Smoking also increases the risk of cancers of the throat, mouth, esophagus, larynx (voice box), pancreas, kidney, bladder, stomach, and cervix, as well as acute myeloidleukemia (US Department of Health and Human Services, 2004; National Cancer Institute, 2008). Smokers are up to six times more likely to suffer from heart attacks compared with nonsmoking counterparts (US Department of Health and Human Services, 2005). Smoking is also harmful during pregnancy. A pregnant woman who smokes has a greater likelihood of having her baby born prematurely and with an abnormally low birth weight (US Department of Health and Human Services, 2004). Furthermore, smoking during or after pregnancy increases an infant's risk of death from Sudden Infant Death Syndrome (SIDS).

Quitting smoking can lead to substantial health benefits, both immediate and long term. Within a few hours of quitting smoking the level of carbon monoxide in the blood (which reduces the blood's ability to carry oxygen) begins to decline. Within a few weeks there are additional improvements, including improved circulation and less coughing and wheezing, and within months there are already marked improvements in lung function (Peto et al., 2000). The long term benefits include but are not limited to reduced risk of cancer, heart disease and lung disease (Doll et al., 2004; McBride & Ostroff, 2003). Individuals who quit smoking, regardless of their age, live longer than those who continue to smoke. That said, quitting smoking is particularly beneficial for younger people. Research indicates that quitting at around age 30 can reduce the chance of dying from smoking related diseases by more than 90% (Doll et al., 2004; McBride & Ostroff, 2003).

Despite the significant benefits to quitting smoking, and the fact that most smokers desire to quit, smokers often fail in their attempts to achieve long term abstinence. Because of the extremely harmful implications of smoking it is essential that researchers continue to seek out explanations that may assist in the prevention and treatment of smoking. Research about, and models of, self-control may offer a unique perspective to explain the difficulties that smokers may face, the various barriers that may explain smokers reluctance to quit, and most importantly provide potential means to reduce smoking initiation and relapse. The ability to exert self-control is critical for successful smoking cessation, because smoking becomes a strongly engrained and automatic pattern of behavior and someone who typically smokes and wishes to quit must exert self-control in order resist that urge and break the habit (Brown, 1998; Tiffany, 1990).

Dieting and Weight Managment

Healthy eating is associated with reduced risk for many diseases, including cardiovascular disease, cancer, and stroke (U.S. Department of Health and Human

Services, 2001). In the US most young people are not following the recommendations set forth in the Dietary Guidelines for Americans (U.S. Department of Agriculture, 1998). Specifically, among U.S. youth aged 6-19, 67% exceed dietary guidelines recommendations for fat intake, and 72% exceed recommendations for saturated fat intake. In 2007, only 21.4% of high school students reported eating fruits and vegetables five or more times daily (when fried potatoes and potato chips are excluded) during the past 7 days (Centers for Disease Control, 2008).

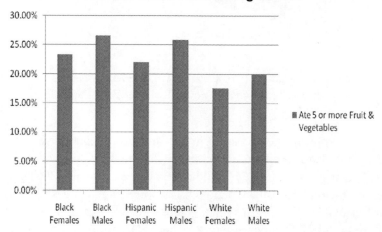

Figure 1. Students (Grades 9-12) who ate the recommended servings of fruit and vegetables (data based on the Youth Risk Behavior Surveillance (Centers for Disease Control, 2008).

Among adults age 20 and over in the United States, 67% are overweight or obese and obesity continues to be a significant public health concern in the U.S. and around the world (National Task Force on the Prevention and Treatment of Obesity, 2000; Flegal et al., 2005; Ogden, Yanovski, Carroll & Flegal, 2007; Ogden, Carroll, McDowell & Flegal, 2007). A growing body of literature suggests that obese individuals face negative attitudes and even overt discrimination from others. Biases towards obese individuals has been demonstrated among children, the general public and even healthcare providers specializing in obesity treatment (Brownell, Puhl, Schwartz & Rudd, 2005; Schwartz, Chambliss, Brownell, Blair & Billington, 2003; Teachman & Brownell, 2001).

Ate Less to Lose Weight

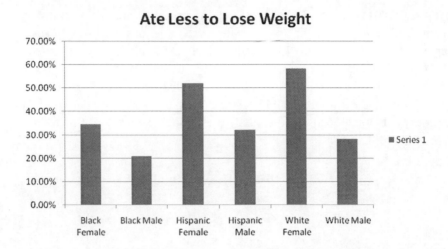

Figure 2. Percentages of Students (Grades 9-12) who ate less or changed types of food in order to lose weight: data based on the Youth Risk Behavior Surveillance (Centers for Disease Control, 2008).

It is not surprising, given the prejudice and discrimination that overweight and obese individuals face that most overweight and obese women desire to lose weight, (Anderson et al., 2002); Wildes, Emery & Simons, 2001). However, regardless of their actual weight, many U.S. women and men desire to weigh less (Maynard et al., 2006).Women typically view themselves as heavier than they actually are and desire a thinner figure (Change & Christakis, 2003; Mossavar-Rahmani, Pelto, Ferris & Allen, 1996) with more reporting dissatisfaction with their bodies than men in the same BMI category (Anderson, Janes, Ziemer & Phillips, 1997). Another study revealed that about 50% of adolescent girls in a rural area reported weight concerns and dieting behavior, and that this increased with age among the sample of 8-17 year olds (Packard & Krogstrand, 2002). According to the Youth Risk Behavior Surveillance (Centers for Disease Control, 2008) in the month prior to completing the survey 40.6% of students nationwide had eaten less food, fewer calories, or low-fat foods to lose weight or to keep from gaining weight. This was particularly high among female compared with male students (See Figure 2). (Centers for Disease Control and Prevention, 2008).

Self-control, the ability to delay immediate gratification, is intricately linked with and is critical to the ability to successfully manage nutritional intake (Baumeister, Heatherton & Tice, 1994). People who wish to limit their consumption of certain foods (e.g., high fat foods) must exert self-control in order to do so. For example, an individual who typically has a high fat desert after dinner would have to exert self-control and delay their immediate gratification (i.e., their desire for a sweet desert) if their ultimate goal was to lose weight. In Western society obese individuals are considered to have weak will, lack self-discipline and engage in hedonistic self-indulgence, whereas a slim figure reflects control and the ability to transcend immediate desires of the flesh (e.g., Murray, 2005; Lupton, 1996). Problem eating, specifically binge eating, has been associated with poor general self-control for both genders (Ricciardelli et al., 2001). Regulating eating behavior is also dependent on an underlying biological process operated by internal factors such as hunger and satiety. However, self-control is necessary for intentional attempts to manage eating, for example for an individual who is attempting to lose weight (Herman & Polivy, 2004). Inhibiting the desire to eat unhealthy foods may be particularly difficult when faced with a situation that is associated with food cues and may trigger the desire to eat the foods, such as the sight and smell of cookies in a bakery.

Dieting, Smoking and Self-control

The previous sections have established the vital role self-control plays in attempts to manage weight and control smoking. It is important to note, that many times and for many different reasons, individuals will attempt to manage both of these demands concurrently. For example, although most people who quit smoking will gain less than 10 pounds (Willamson et al., 1991), concern about weight gain is a perceived barrier to cessation in smokers (Klesges & Klesges, 1988; Pomerleau, Zucker & Stewart, 2001; Russ, Fonseca, Peterson, Blackman & Robbins,). The apprehension regarding weight gain is particularly strong among women (Jeffery et al., 2000). Smokers weigh less on average than non-smokers (Froom et al., 1998) and the majority of women and men do gain some weight with tobacco cessation (Fiore et al., 2008). Concerns regarding cessation related weight gain may lead individuals who are attempting to quit smoking to begin weight management efforts simultaneously. Women and restrained eaters are especially likely to believe that smoking helps to control weight (McKee et al., 2006).

Although dieting may be an intuitive decision for individuals who want to quit smoking but are worried about gaining weight, current treatments strongly advocate *against* dieting during this critical time. Tobacco treatment guidelines strongly encourage clinicians to advise smokers to concentrate on smoking cessation rather than dieting until they are confident that they will not relapse (Fiore et al., 2008). Clinicians are advised to use statements such as: "Try to put your concerns about weight gain on the back burner. You are most likely to be successful if you first try to quit smoking, and then later take steps to reduce your weight. Tackle one problem at a time!...". Research with animals supports these clinical guidelines and demonstrates that caloric restriction increases the self-administration of drugs, including opiates and stimulants (Carr, 2007; Carroll, Campbell & Heideman, 2001; Carroll, Stotz, Kliner & Meisch, 1984). Similarly, studies with humans show that dieting may undermine attempts to quit smoking (Cheskin, Hess, Henningfield & Gorelick, 2005; Hall, Tunstall, Vila & Duffy, 1992) or increase smoking behavior in individuals not attempting to quit. For example, one study found that participants who received either an individualized diet plan intervention or general nutritional educational treatment combined with smoking cessation counseling had higher smoking rates at follow ups compared with a group receiving only smoking cessation counseling (Hall et al., 1992). Conversely, evidence suggests that weight gain or increased intake of sugar may actually predict long-term smoking abstinence (Borrelli, Papandonatos, Spring, Hitsman & Niaura, 2004; Hall, Ginsberg & Jones, 1986; Ockene et al., 2000). In the two studies presented in this chapter, the self-control strength model is applied in order to more closely examine the link between food restriction and smoking behavior. Specifically, it tests whether resisting the temptation to eat from a plate of tempting foods (e.g., cookies, candy) will affect subsequent smoking behavior.

Self-control Replenishment

The previous sections have demonstrated that exertion of self-control depletes self-control strength or resources, and is detrimental to subsequent self-control performance. However, it is reasonable to assume that this effect is only temporary in nature. If self-control depletion was permanent, every act of self-control would permanently deplete people's resources and they would spiral down to a progressive inability to exert any self-control over their behavior. Somehow, the resources expended during various acts of self-control are replenished. To date, less is known about replenishment of self-control and volitional resources than the depletion of this resource. One potential way of replenishing self-control strength is rest or sleep (Baumeister, Muraven & Tice, 1998). Long periods of not exerting self-control will allow for the natural recovery of self-control strength, thereby neutralizing any effects of exerting self-control. This nicely follows the perspective of self-control as a muscle, which may recover physical capacity after a period of rest.

A variety of experiments support the replenishing role of rest in self-regulation ability (for a review see Baumeister, Heatherton & Tice, 1994). Studies demonstrate that people regulate themselves better in the morning after a good night's rest, than later in the day when they are tired. For instance, in one study people abstaining from smoking performed better on a test of attention regulation after sleeping than when they were tired (Parrott, Garnham, Wesnes & Pincock, 1996). In another study dieters were shown to be more likely to relapse and break their diets at night, after they had presumably been self-regulating their nutritional intake all day (Grilo, Shiffman & Wing, 1989). Similarly, bulimics were shown to binge mainly later in the day, when they were tired and presumably depleted in terms of self-control resources (Johnson & Lason, 1982). Stunkard (1959) described a phenomenon of "night bingeing", a pattern in which obese people would eat large amounts of food late in the evening or at night. Exerting self-control can be difficult, whether an individual is attempting to quit smoking, resist tempting foods, or perform any other volitional task. These studies demonstrate that these tasks of self-control may seem more difficult towards the end of the day, when resources become depleted, than after a period of sleep or rest. Thus, sleep or rest may replenish the capacity for self-regulation. Sleep deprivation appears to be linked to impairments in the capacity to self -regulate, further supporting the importance of sleep in replenishing self-control strength. Sleep deprivation has been shown to lead to reduced motivational strength, worsening moods, and decreased ability to concentrate (Mikulincer, Babkoff, Caspy & Sing, 1989). Likewise, sleep deprivation has been shown to cause apathy, anxiousness, and irritability (Murray, 1965). Since the occurrence of sleep deprivation is relatively rare and extreme in intensity, other studies have investigated more common and milder sleep disturbances. These studies uncovered a pattern similar to that of sleep deprivation. Specifically, they found that sleep disturbances such as insomnia and sleep apnea were linked to psychopathologies such as depression, anxiety, and obsessive-compulsive disorder (Coursey, Buchsbaum & Frankel, 1975; Monroe, 1967; Monroe & Marks, 1977). Thus, disturbances or more extreme deprivation of sleep may undermine people's ability to exert self-control.

Another technique that has been used to increase self-control capacity is to progressively increase self-control strength through a variety of self-regulatory exercises or practice. This fits nicely with the metaphor of self-control as a muscle or strength. Although a muscle may become fatigued after physical exercise, it can also be strengthened over time through regular exercise. In the same way studies have demonstrated that self-control strength increases after gradual and regular self-regulatory exercise. In a series of three longitudinal studies researchers attempted to increase participants self-control strength by having them practice self-regulatory exercises over a two week period. For example, in one study participants were instructed to try to use their nondominant hand, and in another they were to inhibit or refrain from cursing. At the end of the two week period participants performed better on self-control measures involving suppression and persistence (Gailliot, Plant, Butz & Baumeister, 2007). Another study instructed participants to practice self-regulatory tasks (involving a program of regular physical exercise) and tracked their self-control performance at baseline and at monthly intervals over a 4-month period. Participants in the experimental condition, who had practiced self-control by following a regular exercise regimen, showed significant improvements in various domains of self-control. This included decreases in alcohol and caffeine consumption, smoking, greater emotional control, and better performance on a laboratory measure of self-control involving a visual tracking task (Oaten & Cheng, 2006).

Another theory, which is tested in the second experiment presented in this chapter, suggests that positive emotion or mood may replenish self-control strength. A recent series of laboratory experiments demonstrated that participants who were depleted by exerting self-control on an initial task of resisting temptation and were then induced with positive affect were able to perform better than those who were not induced with positive affect (Tice, Baumeister, Shmueli & Muraven, 2007). That is, positive emotions were able to counteract the effects of self-control depletion. For instance, participants who were depleted by having to exert self-control to resist the urge to eat tempting sweets and then watched a humorous comedy clip performed better on a subsequent self-control persistence measure compared to those who watched a neutral documentary film. It seems that the positive emotions induced by watching the humorous clip replenished participants' self-control strength. In order to reduce alternative explanations based on particular procedures and to increase generalizability, converging evidence was sought out by using different procedures and measures in a series of four separate experiments (Tice, Baumeister, Shmueli & Muraven, 2007). Self-control depletion was manipulated by having participants breaking an acquired habit, suppressing thoughts, or resisting tempting sweets. Positive mood was manipulated by having participants watch a funny film clip, or giving them a gift of candy. Finally, subsequent self-control performance was measured by two different persistence tasks, a frustration tolerance test, or making oneself drink a healthy but bad tasting beverage. Additionally, in order to ascertain that it was specifically *positive* affect one study induced a sad mood as well. Participants in the sad condition did not show any recovery from the effects of depletion.

Previous research on emotions supports the idea that positive emotions may be beneficial and help replenish the self's resources. Positive moods may influence people's appraisal of their ability to withstand negative events and information, or alter the weight of immediate costs relative to long-term gains

(Trope & Pomerantz, 1998; Trope & Neter, 1994). A series of studies demonstrated that positive experiences made the trade-off between immediate emotional loss and long-term informational benefits more appealing. Specifically, participants who experienced (or even thought about), positive experiences were more willing to accept negative feedback about their performance on an unrelated task. In the short term negative feedback may not be particularly desirable, but in the long term it would provide pertinent information about the self, and the opportunity for self-improvement or avoidance of certain situations. From this perspective, positive moods are converted into usable resources and thus replenish individuals' capacity.

The idea that positive affect may be beneficial to self-control efforts is supported by studies showing positive emotions can help undo many of the harmful physiological effects caused by negative experiences and even create an 'upwards spiral' of positive affect (e.g., Fredrickson, 2001; Fredrickson & Levenson, 1998; Fredrickson, Mancuso, Branigan & Tugade, 2000, Fredrickson & Joiner, 2002). For example, after watching a fear inducing film clip participants were shown additional clips that induced either positive, neutral, or negative emotions. Those who viewed the film that induced positive emotions showed the most rapid return to the pre-film levels of cardiovascular activation. Studies have also shown that positive affect has a strong association with health outcomes over time. For instance, positive affect was predictive of risk of stroke in a sample of over 2,400 adults (Ostir et al., 2001) and was also linked to mortality in an HIV+ sample of 1,043 single men (Moskowitz, 2003).

Previous research has also shown a specific association between positive affect and cigarette smoking. In a longitudinal treatment study higher levels of baseline positive mood were associated with an increased probability of abstinence at the10-week follow up (Doran et al., 2006). Studies have also demonstrated a link between positive affect and smoking craving and relapse. In one study individuals with chronically low positive mood showed greater increases in craving 24 hours after nicotine withdrawal (Cook et al., 2004). Diminished positive mood after quitting was also shown to predict increased likelihood of relapse (al'Absi et al., 2004).

Similarly, literature on negative emotions demonstrates an opposite relationship with smoking outcomes. Negative affect has shown to be detrimental, and to play a role in both overeating and smoking. Smokers who experience negative emotional arousal tend to use cigarettes to enhance their mood (Pomerleau & Pomerleau, 1987). When individuals who restrain their eating are distressed they typically eat more (e.g., Rutledge & Linden, 1998). Studies using Ecological Momentary Assessment (EMA), a sophisticated daily diary method, have shown that rapid increases in negative affect were associated with smoking relapse (Shiffman & Waters, 2004). Similarly, many relapses are marked by intense negative affect and by increases in negative affect in the hours preceding relapse (Shiffman, 2005).

The Present Experiments

The previous section presented results from numerous psychological experiments that illustrated that inhibiting a behavior or engaging in other acts that require self-control significantly reduces individuals' subsequent ability to regulate their behavior. In a series of two controlled randomized laboratory experiments this self-control depletion effect has been applied to smoking behavior to test whether depleting self-control in an initial task would leave people more vulnerable to smoke subsequently. Thus, one of the main strengths of the currents studies are their extension of the literature by applying the self-control paradigm to understanding the competing demands of tobacco use and dietary restraint. The second study extends the self-control paradigm to include not only self-control depletion, but also self-control replenishment using positive emotions. That is, it tests whether being induced with a positive mood can counteract the effect of self-control depletion, and cause people to be less likely to smoke subsequently.

Experiment 1

Experiment 1 applied the self-control depletion model in order to examine the relationship between eating and smoking behavior. The goal of this first study was to test whether engaging in an initial task of inhibition (resisting eating sweets) would affect subsequent smoking behavior. The sample consisted of 101 smokers (54 male, 41 female, and 6 transgender) from the San Francisco Bay Area. Recruitment was done using flyers and ads posted in the community section of 'Craigslist'. Inclusion criteria included legal smoking age (> 18 years old) and smoking a minimum of one cigarette a week. Participants were told that the purpose of the experiment was to test whether smokers are able to resist the urge to eat from different foods when exposed to their sight and smell. Participants were tested individually in sessions lasting approximately one hour (M =49.3 minutes, SD =8.62). They completed a set of pre-experimental measures including a demographic measure, measures assessing their smoking dependency, frequency, and readiness to quit, and their general liking for a desire for sweets and vegetables at that moment. Participants also completed a baseline measure of expired air carbon monoxide (CO) using a Bedfont Smokerlyzer to evaluate participants' expired CO levels and obtain a specific personalized reading of the level of carbon monoxide in participants' lungs in measures of parts per million (range 1-80ppm). This measure has been used in clinical tobacco cessation trials and provides an unbiased and reliable indicator of recent smoking (e.g., Irving, Clark, Crombie & Smith, 1988; Jarvis, Belcher, Vesey & Hutchinson, 1986).

Participants then completed the cue exposure phase of the study, which followed standard cue reactivity assessments used in similar studies (Monti et al., 1993). They were presented with a large plate of either tempting sweets (high fat sweets such as brownies and chocolate chip cookies) or vegetables (radishes and broccoli), according to a predetermined randomized order. Participants were instructed to resist eating from the food while the experimenter left the room. During that time, they were to follow a 5-minute prerecorded tape with a series of bell rings (the duration between rings varied, $M = 15$ seconds). Every time the bell rang they were to lift the plate and smell the food for a few seconds before putting it back down. They also were instructed to think about the temptation involved in resisting the food, although it was emphasized that they should try not to actually eat any of it. This task was designed to simulate the primary self-control demand associated with dieting, which is the inhibiting the urge to eat tempting foods. We operationalized dieting in this way in order to be able to capture this self-control demand in a laboratory setting.

Following the food cue exposure task, participants completed a manipulation check assessing their experience of the task. They were then given a 10-minute break, before continuing with the rest of the study. The participants were asked to leave the room so that the experimenter could prepare the next phase of the study, but given no restrictions as to what to do during the break. The room in which the study took place was located on the ground floor of a medical center with a large lobby and easy access to the sidewalk outside. Thus, participants' potential alternatives were to remain in the indoor lobby waiting area or go outside. Unbeknownst to the participants, whether or not they chose to smoke during the break was the primary dependent variable of the study. Upon their return to the testing room, the experimenter took another expired CO sample reading, which would be used in conjunction with the baseline reading in order to determine smoking during the break in the study. Finally, participants were debriefed, paid for their time and effort, and thanked for their participation.

Results showed that participants who resisted the cookies and sweets rated the task as more tempting than those who resisted the plate of raw vegetables. This suggests that the food manipulation, designed to create a task that depleted self-control strength by demanding exertion of self-control to inhibit desires, succeeded. Furthermore, as predicted, participants who resisted the sweets were more likely to smoke during the break in the study, as indicated by corresponding changes in their CO levels (See Figure 3). Regardless of their experimental condition, participants who were more dependent smokers, and those who smoked more cigarettes on average per day, were more likely to smoke during the break. Thus, the results were consistent with the self-control strength model showing a pattern of an initial self-control task impairing a subsequent one.

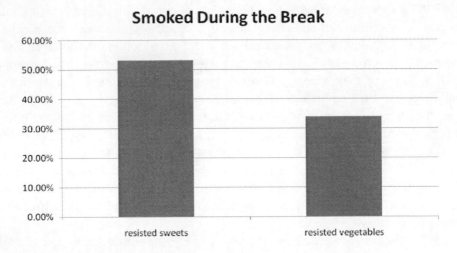

Figure 3. Percent of Participants Who Smoked During the Break.

Experiment 2

The second study had two goals. The first goal was to replicate the results of Experiment 1, and support the prediction that an initial task of resisting tempting foods would impair subsequent self-regulatory performance, specifically controlling smoking. The second goal of the study was to extend the self-control model by testing a brief affective intervention, specifically whether a positive affect induction could counteract the detrimental effects of self-control depletion. As such, the procedures were identical, other than the inclusion of a mood manipulation.

After completing the self-control depletion task (i.e., resisting either a plate of tempting sweets or less tempting raw vegetables) participants were randomly assigned to a positive or neutral affect induction group. Those in the experimental group were induced with positive emotion either by viewing a humorous video clip of the comic Robin Williams or writing about a positive experience that they had. Participants in the control group watched a 5-minute clip from a neutral film about building bridges or wrote a detailed description of a room in their house. Participants were all then given a 10-minute break, before continuing with the rest of the study. Upon their return to the room, participants were asked to provide another expired CO sample. Finally, participants were debriefed, paid for their time and effort, and thanked for their participation in the study.

A total of 200 participants (122 male, 70 female, 8 transgender) were included in the study. As in the previous experiment, participants who resisted the cookies reported higher temptation compared with those who resisted the vegetables, indicating the success of the self-control depletion manipulation. Regarding the primary variable of interest in the study: whether participants smoked during the 10-minute study break, results were again consistent with those of the previous study. Specifically, participants who resisted sweets were more likely to smoke during the break compared with those who resisted vegetables. Furthermore, a logistic interaction model indicated an interaction between self-control depletion (food: sweets or vegetables) and affect (valence: positive or neutral). Among participants randomly assigned to cookies, those who were in the neutral condition were more likely to smoke than those in the positive affect group. Regardless of experimental condition, a logistic regression analysis indicated a statistically significant relationship between smoking dependency and likelihood of smoking during the break. That is, smokers who were more heavily dependent were more likely to smoke during the break in the study. Additionally, smokers in the preparation stage of change (who were planning to quit smoking within the next 30 days) were less likely to smoke during the break in the study.

The results of the study were consistent with the self-control strength model showing a pattern of an initial self-control task (resisting tempting foods) impairing a subsequent one (smoking). Furthermore, positive emotions appeared to counteract this detrimental effect of self-control depletion. Depleted smokers who experienced the positive mood induction were as unlikely to smoke during the break as smokers who were not depleted in the first place (who resisted vegetables). That is, the positive affect was equivalent to a return to a pre-depleted state, suggesting important implications for tobacco cessation interventions.

Conclusion

Self-control and Smoking

In the two studies presented in this chapter the self-control depletion and self-control replenishment paradigms were tested to investigate whether participants' smoking behavior was consistent with the model's predictions. The results of both controlled randomized experiments supported these paradigms and demonstrate that self-control is in fact a mechanism underlying smoking behavior. Specifically, in both studies, smokers who resisted tempting sweets were more likely to smoke a cigarette during a break in the study, compared to smokers who resisted raw vegetables. That is, participants whose self-control strength was depleted by the initial task of inhibition were more likely to smoke when given an opportunity to do so in the ensuing break. This pattern of one self-control task impairing a subsequent self-control related task is consistent with previous work on successive attempts at self-regulation (e.g., Baumeister et al., 1998).

The study findings offer a unique explanation for how fighting against a temptation affects subsequent smoking behavior. Exposure to a tempting substance (e.g., cookies and sweets in the current studies) triggers an automatic inclination to eat from the food, and requires self-control to inhibit that behavior. This exertion of self-control in order to inhibit the behavior may deplete an individual's self-control strength or capacity and put him or her at risk for subsequently increasing habitual behaviors such as smoking.

The second study extended the findings by investigating the possibility of increasing depleted self-control strength. The rationale for the replenish paradigm is that if depletion of self-control strength is responsible for difficulties controlling smoking, then increasing or replenishing self-control strength should help to prevent that effect. Results of Experiment 2 supported that assertion and demonstrated that positive affect was able to counteract the effect of self-control depletion. Specifically, smokers who were depleted (by resisting tempting sweets) but then experienced a positive affect induction were less likely to smoke compared with depleted smokers who received a neutral affect induction.

Smoking and Dieting

The findings of the current studies have implications for understanding the issue of co-occurring dietary restraint and smoking cessation. Many instances may involve simultaneous restriction of certain foods and tobacco use simultaneously, such as health restrictions following an illness or medical event, pregnancy, or simply a desire to change both of these issues (e.g., as a New Year's resolution). The present findings suggest that since inhibiting one behavior may impair the other, it would be best to focus on only one behavior initially. This is consistent with research that suggests that intervention success is greater when behavior changes are initiated sequentially rather than simultaneously (Spring et al., 2004). In the case of smoking cessation and dieting the recommendation would generally be to focus on smoking cessation, since the consequences of smoking may be more harmful. Thus, the findings support current tobacco treatment clinical practice guidelines, which discourage concurrent dieting during cessation attempts (Fiore et al., 2008) and recommend that smokers focus solely on smoking cessation and ignore weight management until they are confident that they will not relapse. Other experiments also have supported the idea that dieting during smoking cessation may increase the risk of relapse, and longitudinal studies have demonstrated that weight gain during early abstinence is predictive of long-term abstinence rather than relapse (Borrelli, Papandonatos, Spring, Hitsman & Niaura, 2004; Hall, Ginsberg & Jones, 1986; Ockene et al., 2000). The results of the current studies suggest dietary restraint may deplete an individual's self-control resources, leaving them more prone to fail on the additional self-control task of smoking inhibition.

Ultimately, this could lay the foundation for an intervention for individuals attempting to manage the co-occurring regulatory demands of dietary restraint and smoking cessation and provide useful information for clinicians. For instance, a measure assessing the environmental cues people experience in their daily lives (e.g., tempting foods in their home, quantity of food advertisements individuals perceive) may assist in explaining and increasing patients' successful smoking cessation.

The findings from Experiment 2 are also particularly useful in terms of their potential for informing treatment interventions for smokers who are trying to quit and who are concerned with post-cessation weight gain. The *Self-Control Strength Model* and these current studies in particular, demonstrate that exerting self-control by dieting may impair attempts to control smoking, but positive emotions may be able to counteract this effect. Therefore, incorporating a positive affect intervention as part of treatment may increase or replenish self-control strength and help to reduce the effect of self-control depletion. These results support previous interventions that have incorporated mood management in smoking cessation treatment (Hall et al., 1998; Hall et al., 2002; Hall et al., 2004, Muñoz et al., 2006). For instance, models of addictive behavior have long since recognized the link between negative affect and substance use (Shiffman et al., 2007).

Extensive work in the field of stress and coping literature (e.g., Wills & Shiffman, 1985) also suggests that one of the roles of various substances is to regulate emotions and cope with stress. Other research has demonstrated that many smokers believe that smoking will help alleviate negative moods (Brandon & Baker, 1991) and abstinence from smoking is associated with withdrawal symptoms including negative affect states such as anxiety and irritability (al'Absi et al., 2002; Hughes, Higgins & Hatsukami, 1990). These negative affect withdrawal symptoms have been shown to be determinants of early smoking lapse (Kenford et al. 2002; Piasecki et al., 2003), which makes mood management of primary importance during smoking cessation treatment. Experiment 2 extends the recognized link between emotions and smoking by specifically examining positive affect rather than negative affect as well as offering an innovative theoretical explanation for this demonstrated link.

Unanswered Questions and Future Directions

Although the experimental findings presented offer promising results that extend the current literature involving self-control and smoking behavior, there are several shortcomings that should be addressed. One key issue involves logistical concerns. In both studies presented in this chapter, there were some participants who reported a desire to smoke (during the break in the study) but who were unable to do so for lack of cigarettes. Providing participants with cigarettes was not considered in the study design. Another factor that may have contributed to the fact that some participants did not carry cigarettes with them on the day of testing was smoking frequency. Our inclusion criteria were relatively broad regarding average cigarettes smoked per week, and as a result part of our sample was comprised of light smokers (who smoked less than 10 cigarettes a day). These social or light smokers may have been less likely to carry cigarettes with them on a daily basis, and thus did not have them on the day of testing. In future studies it would be interesting and important to replicate the findings with a sample of heavy smokers. Additionally, it may be useful in future studies to instruct participants to refrain from eating or smoking for several hours before the experiment in order to strengthen the manipulations.

Another limitation of the two study designs were methodological in nature, and included the use of a between-subjects design and the potential priming of smoking behavior. Although the use of a within-subjects design has clear benefits, it was necessary to ensure participants naiveté regarding the smoking outcome measured immediately after the break in the study. Participants in the study were unaware that their choice to smoke during the break was a significant variable in the study or that it would be documented or measured, until they returned to the lab and were asked to provide a second CO reading. If we had used a within-subject design, with an additional break after a second food cue exposure, their knowledge of the approaching smoking verification may have interfered with their true desired behavior during the break. In future studies it may be beneficial to conduct the study in an area where it is possible to unobtrusively view or videotape participants during the break, thus allowing for a within-subjects design.

There are several important implications that arise from the study findings. Most prominently, the findings have implications for understanding co-occurring self-regulation demands in daily life and may be particularly relevant for multiple health behavior change interventions with smokers. Research has shown that tobacco users, in particular, tend to have poor health behavior profiles, with the overwhelming majority of smokers (92%) engaging in at least one additional risk behavior such as other substance use or having a sedentary lifestyle (Fine, Philogene, Gramling, Coups & Sinha, 2004; Klesges, Eck, Isbell, Fulliton & Hanson, 1990). The results of the current studies suggest that innovative interventions designed to address multiple risks in tobacco users may need to consider staggering or delaying secondary change efforts in order to reduce the amount of self-control demand an individual must deal with at the same time. Future studies may wish to examine additional instances of co-occurring self-regulation demands, such as desires to quit concurrent smoking and alcohol use or other substance use.

Results of Experiment 2 demonstrate that positive affect may be able to counteract the effects of self-control depletion occurring from the exertion of self-control on more than one task simultaneously. Thus, treatments designed to incorporate a positive affect intervention may be particularly useful for individuals who are attempting to deal with concurrent self-control demands, such as smoking cessation and alcohol reduction. The benefits of positive emotions may be applied beyond treatment, to the general public as well. Any situation which involves two demanding self-regulatory tasks may be relevant. For example, students who are trying to maintain healthy habits (e.g., diet, exercise) while studying for demanding final exams may also benefit from experiences of positive emotion during these stressful times.

Concluding Remarks

To conclude, self-control may be fundamental to understanding smoking behavior. In particular, these studies applied an innovative model of self-control, the *self-control strength model*, which demonstrates that momentary self-control strength or capacity is intricately linked with the likelihood of smoking. This conceptualization of self-control, as a fluctuating state dependent on previous exertion of self-control, is a unique and promising new view that extends previous research on trait self-control and smoking behavior. It is also particularly useful for understanding the difficulties associated with multiple health behavior change, such as smoking cessation and dieting. Moreover, the model incorporates the possibility of replenishing or increasing self-control ability, which may provide a foundation for treatment interventions designed to reduce the incidence of tobacco use.

References

al'Absi, M., Hatsukami, D., Davis, G. L. & Wittmers, L. E. (2004). Prospective examination of effects of smoking abstinence on cortisol and withdrawal symptoms as predictors of early smoking relapse. *Drug & Alcohol Dependence, 73*, 267-278.

al'Absi, M., Amunrud, T. & Wittmers, L. E. (2002). Psychophysiological effects of abstinence and behavioral challenges in habitual smokers, *Pharmacology. Biochemistry, and Behavior, 72*, 707–716.

Anderson L. A., Eyler, A. A., Galuska, D. A., Brown, D. R. & Brownson, R. C. (2002). Relationship of satisfaction with body size and trying to lose weight in a national survey of overweight and obese women aged 40 and older, United States. *Preventive Medicine, 35*, 390–396

Anderson, L. A., Janes, G. R., Ziemer, D. C. & Phillips, L. S. (1997). Diabetes in urban African Americans. Body image, satisfaction with size, and weight change attempts. *Diabetes Education, 23*, 301–308.

Ayduk, O., Mendoza-Denton, R., Mischel, W., Downey, G., Peake, P. & Rodriguez, M. (2000). Regulating the interpersonal self: Strategic self-regulation for coping with rejection sensitivity. *Journal of Personality and Social Psychology, 79*, 776-792.

Barkley, R. A. (1997). Behavioral inhibition, sustained attention, and executive functions: Constructing a unifying theory of ADHD. *Psychological Bulletin, 121*, 65-94.

Baumeister, R. F. (2005). *The cultural animal: Human nature, meaning, and social life.* New York: Oxford University Press.

Baumeister, R. F., Heatherton, T. F. & Tice, D. M. (1994). *Losing control: How and why people fail at self-regulation.* San Diego, CA: Academic Press.

Baumeister, R. F., Bratslavsky, E., Muraven, M. & Tice, D. M. (1998). Ego-depletion: Is the active self a limited resource? *Journal of Personality and Social Psychology,* **74**, 1252-1265.

Baumeister, R. F., Muraven, M. & Tice, D. M. (2000). Ego depletion: A resource model of volition, self-regulation, and controlled processing. *Social Cognition,* **18**, 130-150.

Borrelli, B., Papandonatos, G., Spring, B., Hitsman, B. & Niaura, R. (2004). Experimenter-defined quit dates for smoking cessation: adherence improves outcomes for women but not for men. *Addiction,* **99**(3), 378-385.

Brandon, T. H. & Baker, T. B. (1991). The smoking consequences questionnaire: The subjective expected utility of smoking in college students. *Psychological Assessment,* **3**, 484–491

Brandon, J. E., Oescher, J. & Loftin, J. M. (1990). The self-control questionnaire: An assessment. *Health Values,* **14**, 3–9.

Brown, J. M. (1998). Self-regulation and the addictive behaviors. In W. R. Miller & N. Heather (Eds.), *Treating addictive behaviors* (2nd ed., 61-73). New York: Plenum.

Brownell, K. D., Puhl, R. M., Schwartz, M. B. & Rudd, L. (2005). *Weight bias: Nature, consequences, and remedies.* New York, NY, US: Guilford Publications, 320.

Carr, K. D. (2007). Chronic food restriction: enhancing effects on drug reward and striatal cell signaling. *Physiology & Behavior,* **91**, 459-72.

Carroll, M. E., Campbell, U. C. & Heidman, P. (2001). Ketoconazole suppresses food restriction-induced increases in heroin self-administration in rats: Sex differences. *Experimental and Clinical Psychopharmacology,* **9**, 307-316.

Carroll, M. E., Stotz, D. C., Kliner, D. J. & Meisch, R. A. (1984). Self-administration of orally-delivered methohexital in rhesus monkeys with phencyclidine or pentobarbital histories: Effects of food deprivation and satiation. *Pharmacology, Biochemistry and Behavior,* **20**(1), 145-151.

Centers for Disease Control and Prevention (2007). Cigarette Smoking Among Adults - United States. *Morbidity and Mortality Weekly Report.*

Centers for Disease Control and Prevention (2008). Smoking-Attributable Mortality, Years of Potential Life Lost, and Productivity Losses—United States, 2000–2004. *Morbidity and Mortality Weekly Report,* **57**(**45**), 1226–8

Centers for Disease Control and Prevention (2006). *2006 National Youth Tobacco Survey and Key Prevalence Indicators*.

Centers for Disease Control and Prevention (2008). Youth Risk Behavior Surveillance- United States, 2007. *Morbidity & Mortality Weekly Report* **57**, (SS-05), 1–131.

Centers for Disease Control and Prevention (2005). Annual Smoking-Attributable Mortality, Years of Potential Life Lost, and Productivity Losses--United States, 1997–2001. *Morbidity and Mortality Weekly Report 2005;* **54***(25)*, 625–628.

Chang, V. W. & Christakis, N. A. (2003). Self-perception of weight appropriateness in the United States. *American Journal of Preventive Medicine,* **24**, 332–339.

Cheskin, L. J., Hess, J. M., Henningfield, J. & Gorelick, D. A. (2005). Calorie restriction increases cigarette use in adult smokers. *Psychopharmacology,* **179**(2), 430-436.

Cook, J. W., Spring, B., McChargue, D. & Hedeker, D. (2004). Hedonic capacity, cigarette craving, and diminished positive mood. *Nicotine & Tobacco Research,* **6**, 39-47.

Costa, P. T. Jr. & McCrae, R. R. (1992). *Revised NEO Personality Inventory (NEO-PI-R) and NEO Five-Factor Inventory (NEO-FFI) professional manual*. Odessa, FL: Psychological Assessment Resources.

Coursey, R. D., Buchsbaum, M. & Frankel, B. L. (1975). Personality measures and evoked responses in chronic insomniacs. *Journal of Abnormal Psychology,* **84**, 239-249.

Dewall, C. N., Baumeister, R. F., Gailliot, M. T. & Maner, J. K. (2008). Depletion makes the heart grow less helpful: helping as a function of self-regulatory energy and genetic relatedness. *Personality and Social Psychology Bulletin,* **34**(12), 1653-62.

Doll, R., Peto, R., Boreham, J. & Sutherland, I. (2004). Mortality in relation to smoking: 50 years' observations on male British doctors. *British Medical Journal,* **328**(7455), 1519–1527.

Doran, N., Spring, B., Borrelli, B., McChargue, D., Hitsman, B., Niaura, R. & Hedeker, D. (2006). Elevated positive mood: A mixed blessing for abstinence. *Psychology of Addictive Behaviors,* **20**, 36-43.

Fine, L. J., Philogene, G. S., Gramling, R., Coups, E. J. & Sinha, S. (2004). Prevalence of Multiple Chronic Disease Risk Factors: 2001 National Health Interview Survey. *American Journal of Preventive Medicine. Special Issue: Addressing Multiple Behavioral Risk Factors in Primary Care,* **27**(2,Suppl), 18-24.

Fiore, M. C., Jean, C. R., Baker, T. B., Bailey, W. C., Benowitz, N. L, et al. (2008). *Treating Tobacco Use and Dependence: 2008 Update.* Rockville, MD: US Department of Health and Human Services. Public Health Service.

Flegal, K. M., Graubard, B. I., Williamson, D. F. & Gail, M. H. (2005). Excess deaths associated with underweight, overweight, and obesity. *Journal of the American Medical Association,* **293**(15), 1861–7.

Fredrickson, B. L. & Levenson, R. W. (1998). Positive emotions speed recovery from the cardiovascular sequelae of negative emotions. *Cognition and Emotion,* **12**, 191-220.

Fredrickson, B. L. (2001). The role of positive emotions in positive psychology: The broaden-and-build theory of positive emotions. *American Psychologist,* **56**, 218-226.

Fredrickson, B. L. & Joiner, T. (2002). Positive emotions trigger upward spirals toward emotional well-being. *Psychological Science,* **13**, 172-175.

Froom, P., Melamed, S. & Benbassat, J. (1998). Smoking cessation and weight gain. *Journal of Family Practice,* **46**, 460-464.

Gailliot, M. T., Plant, E. A., Butz, D. A. & Baumeister, R. F. (2007). Increasing self-regulatory strength can reduce the depleting effect of suppressing stereotypes, Personality and Social Psychology Bulletin, **33**(2), 281-94.

Gailliot, M. T. & Baumeister, R. F. (2007). Self-regulation and sexual restraint: dispositionally and temporarily poor self-regulatory abilities contribute to failures at restraining sexual behavior. *Personality and Social Psychology Bulletin,* **33**(2), 173-86.

Gosling, S. D., Rentfrow, P. J. & Swann, W. B. Jr. (2003). A very brief measure of the Big Five personality domains. *Journal of Research in Personality,* **37**, 504–528.

Govorun, O. & Payne, K. B. (2006). Ego-depletion and prejudice: separating automatic and controlled components. *Social Cognition,* **24**, 111-136.

Greenwald, A. G., Oakes, M. A. & Hoffman, H. G. (2003). Targets of discrimination: Effects of race on responses to weapon holders. *Journal of Experimental Social Psychology,* **39**, 399-405.

Grilo, C. M., Shiffman, S. & Wing, R. R. (1989). Relapse crises and coping among dieters. *Journal of Consulting and Clinical Psychology,* **57**, 87-92.

Hall, S. M., Tunstall, C. D., Vila, K. L. & Duffy, J. (1992). Weight gain prevention and smoking cessation: Cautionary findings. *American Journal of Public Health,* **82**(6), 799-803.

Hall, S. M., Ginsberg, D. & Jones, R. T. (1986). Smoking cessation and weight gain. *Journal of Consulting and Clinical Psychology,* **54**(3), 342-346.

Hall, S. M., Humfleet, G. L., Reus, V. I., Muñoz, R. F., Hartz, D. T. & Maude-Griffin, R. (2002). Psychological intervention and antidepressant treatment in smoking cessation. *Archives of General Psychiatry, 59*, 940-936.

Hall S. M., Reus, V. I., Muñoz R. F., Sees, K. L., Humfleet, G., Hartz, D. T., Frederick, S & Triffleman E. (1998). Nortriptyline and cognitive-behavioral therapy in the treatment of cigarette smoking. *Archives of General Psychiatry, 55,* 683-690.

Hall, S. M., Humfleet, G. L., Reus, V. I., Muñoz, R. F. & Cullen, J. (2004). Extended nortriptyline and psychological treatment for cigarette smoking. *American Journal of Psychiatry, 161,* 2100-2107.

Hayes, S. C., Gifford, E. V. & Ruckstuhl, L. E. J. (1996). Relational frame theory and executive function: A behavioral analysis. In G. R. Lyon & N. A. Krasnegor (Eds.), *Attention, memory, and executive function* (279-306). Baltimore: Brookes.

Herman, C. P. & Polivy, J. (1975). Anxiety, restraint, and eating behavior. *Journal of Abnormal Psychology, 84,* 666-672.

Herman, C. P. & Polivy, J. (2004). The self-regulation of eating: Theoretical and practical problems. In R. F. Baumeister & K. D. Vohs (Ed), *Handbook of self-regulation: Research, theory, and applications* (492-508). New York, NY, US: Guilford Press.

Hughes, J. R., Higgins, S. T. & Hatsukami, D. (1990). *Effects of abstinence from tobacco: A critical review.* In L. T. Kozlowski, H. M. Annis, H. D. Cappell, F. B.

Glaser, M. S. Goodstat, Y. Israel, et al. (Eds.), *Research advances in alcohol and drug problems* (Vol. 10, 317–398). New York: Plenum.

Irving, J. M., Clark, E. C., Crombie, I. K. & Smith, W.C. (1988). Evaluation of a portable measure of expired-air carbon monoxide. *Preventative Medicine, 17,* 109–115.

Jarvis M. J., Belcher M., Vesey, C. & Hutchinson, D. C. S. Low cost carbon monoxide monitors in smoking assessment. *Thorax 1986; 41,* 886-887.

Johnson, C. & Lason, R. (1982). Bulimia: An analysis of moods and behavior. *Psychosomatic Medicine, 44,* 341-351.

Jeffery, R. W., Hennrikus, D. J., Lando, H. A., Murray, D. M. & Liu, J. W. (2000). Reconciling conflicting findings regarding postcessation weight concerns and success in smoking cessation. *Health Psychology, 19,* 242-246.

Kelly, E. L. & Conley, J. J. (1987). Personality and compatibility: A prospective analysis of marital stability and marital satisfaction. *Journal of Personality and Social Psychology, 52,* 27-40.

Kenford, S. L., Smith, S. S., Wetter, D.W., Jorenby, D. E., Fiore, M. C. & Baker, T. B. (2002) Predicting relapse back to smoking: Contrasting affective and physical models of dependence, *Journal of Consulting and Clinical Psychology* **70**, 216–227

Klesges, R. C., Eck, L. H., Isbell, T. R., Fulliton, W. & Hanson, C. L. (1990). Smoking status: effects on the dietary intake, physical activity, and body fat of adult men. *American Journal of Clinical Nutrition, 1990*, 784-789.

Klesges, R. C. & Klesges, L. M. (1988). Cigarette smoking as a dieting strategy in a university population. *International Journal of Eating Disorders,* 7, 413-419.

Kochanska, G., Murray, K. T. & Harlan, E. T. (2000). Effortful control in early childhood: Continuity and change, antecedents, and implications for social development. *Developmental Psychology,* 36, 220–232.

Lupton, D. (1996). *Food, the body and the self,* Sage, London.

Maynard, L. M., Serdula, M. K., Galuska, D. A., Gillespie, C. & Mokdad, A. H.(2006).Secular trends in desired weight of adults. *International Journal of Obesity,* **30**, 1461.

McBride, C. M & Ostroff, J. S. (2003). Teachable moments for promoting smoking cessation: The context of cancer care and survivorship. *Cancer Control*, **10**(4), 325–333.

McKee, S. A., Nhean, S., Hinson, R. E. & Mase, T. (2006). Smoking for weight control: Effect of priming for body image in female restrained eaters. *Addictive Behaviors*.

Milkulincer, M., Babkoff, H., Caspy, T. & Sing, H. C. (1989). The effects of 72 hours of sleep loss on psychological variables. *British Journal of Psychology,* **80**, 145-162.

Mischel, W. (1974). Cognitive appraisals and transformations in self-control. In B. Weiner (Ed.), *Cognitive views of human motivation* (33-49). New York: Academic Press.

Mischel, W. & Shoda, Y. (1995). A cognitive-affective system theory of personality: Reconceptualizing situations, dispositions, dynamics, and invariance in personality structure. *Psychological Review,* **102**, 246-268.

Mischel, W., Shoda, Y. & Peake, P. (1988). The nature of adolescent competencies predicted by preschool delay of gratification. *Journal of Personality and Social Psychology,* **54**, 687-696.

Mischel, W., Shoda, Y. & Rodriguez, M. L. (1989). Delay of gratification in children. *Science,* **244**, 933-938.

Monroe, L. J. (1967). Psychological and physiological differences between good and poor sleepers. *Journal of Abnormal Psychology,* **72**, 255-264.

Monroe, L. J. & Marks, P. A. (1977). Psychotherapists' descriptions of emotionally disturbed adolescent poor and good sleepers. *Journal of Clinical Psychology, 33*, 263-269.

Monti, P. M., Rohsenow, D. J., Rubonis, A. V., Niaura, R. S, Sirota, A. D., Colby, S. M., & Abrams, D. B. (1993). Alcohol cue reactivity: Effects of detoxification and extended cue exposure. *Journal of Studies on Alcohol, 54*, 235-245.

Moskowitz, J. T. (2003). Positive affect predicts lower risk of AIDS mortality. *Psychosomatic Medicine, 65*, 620-626.

Mossavar-Rahmani, Y., Pelto, G. H., Ferris, A. M. & Allen, L. H. (1996). Determinants of body size perceptions and dieting behavior in a multiethnic group of hospital staff women. *Journal of the American Dietetic Association, 96*, 252–256.

Muñoz, R. F., Lenert, L. L., Delucchi, K., Stoddard, J., Perez, J. E., Penilla, C. & Perez-Stable, E. J. (2006). Toward evidence-based Internet interventions: A Spanish/English Web site for international smoking cessation trials. *Nicotine & Tobacco Research, 8*, 77-87.

Muraven, M. & Baumeister, R. F. (2000). Self-regulation and depletion of limited resources: Does self-control resemble a muscle? *Psychological Bulletin, 126*, 247-259.

Muraven, M., Collins, R. L. & Nienhaus, K. (2002). Self-control and alcohol restraint: An initial application of the self-control strength model. *Psychology of Addictive Behaviors, 16*, 113-120.

Muraven, M., Collins, R. L., Shiffman, S., and Paty, J. A. (2005). Daily fluctuations in self-control demands and alcohol intake. *Psychology of Addictive Behaviors, 19*, 140-147.

Muraven, M., Pogarsky, G. & Shmueli, D. (2006). Self-control depletion and the general theory of crime. *Journal of Quantitative Criminology, 22*, 263-277.

Muraven, M. & Shmueli, D. (2006a). The self-control costs of fighting the temptation to drink, *Psychology of Addictive Behaviors, 20*, 154-160

Muraven, M., Tice, D. M. & Baumeister, R. F. (1998). Self-control as a limited resource: Regulatory depletion patterns. *Journal of Personality and Social Psychology, 74*, 774-789.

Murray, S. (2005). (Un/be) coming out? Rethinking fat politics, *Social Semiotics, 15*, 153–163. Murray, E. J. (1965). *Sleep, dreams, and arousal*. New York: Appleton-Century-Crofts.

National Cancer Institute. *Smoking and Tobacco Control Monograph 10, Health Effects of Exposure to Environmental Tobacco Smoke.* Bethesda, MD: National Cancer Institute, 1999.

National Cancer Institute (2008). Prevention and Cessation of Cigarette Smoking: *Control of Tobacco Use* (PDQ®).

National Toxicology Program (2005). *Report on Carcinogens. Eleventh Edition.* U.S. Department of Health and Human Services, Public Health Service, National Toxicology Program.

National Task Force on the Prevention and Treatment of Obesity: Overweight, obesity, and health risk. *Archives of Internal Medicine,* **160**, 898–904.

Oaten, M. & Cheng, K. (2006). Longitudinal gains in self-regulation from regular physical exercise. *British Journal of Health Psychology,* **11**, 717-733.

Ockene, J. K., Emmons, K. M., Mermelstein, R. J., Perkins, K. A., Bonollo, D. S., Voorhees, C. C., et al. (2000). Relapse and maintenance issues for smoking cessation. *Health Psychology,* **19**(1 Suppl), 17-31.

Ogden, C. L., Yanovski, S. Z., Carroll, M. D. & Flegal, K. M. (2007). The epidemiology of obesity. *Gastroenterology,* **132**(6), 2087–102.

Ogden, C. L., Carroll, M. D., McDowell, M. A. & Flegal, K. M. (2007). *Obesity among adults in the United States— no change since* 2003–2004. NCHS data brief no 1. Hyattsville, MD: National Center for Health Statistics.

Ostir, G. V., Markides, K. S., Black, S. A. & Goodwin, J. S. (2001). Emotional well-being predicts subsequent functional independence and survival. *Journal of the American Geriatrics Society,* **48**, 473-478.

Packard, P. & Krogstrand, K. S. (2002). Half of rural girls aged 8 to 17 years report weight concerns and dietary changes, with both more prevalent with increased age., *Journal of the American Dietetic Association,* **102**(5), 672-7.

Parrott, A. C., Garnham, N. J., Wesnes, K. & Pincock, C. (1996). Cigarette smoking and abstinence: Comparative effects upon cognitive task performance and mood state over 24 hours. *Human Psychopharmacology Clinical and Experimental,* **11**, 391-400.

Peto, R., Darby, S., Deo, H., et al. (2000). Smoking, smoking cessation, and lung cancer in the U.K. since 1950: Combination of national statistics with two case-control studies. *British Medical Journal,* **321**(7257), 323–329.

Piasecki, T. M., Jorenby, D. E., Smith, S. S., Fiore, M. C. & Baker, T. B. (2003). Smoking Withdrawal Dynamics: I. Abstinence Distress in Lapsers and Abtainers. *Journal of Abnormal Psychology* Vol 112, No. 1, 3-13

Pomerleau, C. S. & Pomerleau, O. F. (1987). The effects of a psychological stressor on cigarette smoking and subsequent behavioral and physiological responses. *Psychophysiology,* **24**, 278-285.

Pomerleau, C. S., Zucker, A. N. & Stewart, A. J. (2001). Characterizing concerns about post-cessation weight gain: results from a national survey of women smokers. *Nicotine and Tobacco Research,* **3**(1), 51-60.

Rachlin, H. (2000). *The science of self-control.* Cambridge, MA, US: Harvard
    University Press.

Ricciardelli, L. A., Williams, R. J., Finemore, J. (2001). Restraint as
    misregulation in drinking and eating. *Addictive Behaviors, 26*, 665-675.

Richards, P. S. (1985). Construct validation of the self-control schedule. *Journal
    of Research in Personality, 19,* 208–218.

Rosenbaum, M. (1980). A schedule for assessing self-control behaviors:
    Preliminary findings. *Behavior Therapy, 11,* 109–121.

Rosenbaum, M & Jaffe, Y. (1983). Learned helplessness: the role of individual
    differences in learned resourcefulness. British Journal of Social Psychology,
    22, 215–225.

Russ, C. R., Fonseca, V. P., Peterson, A. L., Blackman, L. R. & Robbins, A. S.
    (2001).Weight gain as a barrier to smoking cessation among military
    personnel. American journal of health promotion. *American Journal of
    Health Psychology, 16*(2), 79-84.

Rutledge, T. & Linden, W. (1998). To eat or not to eat: Affective and
    physiological mechanisms in the stress-eating relationship. *Journal of
    Behavioral Medicine, 21,* 221-240.

Schmeichel, B. J., Vohs, K. D. & Baumeister, R. F. (2003). Intellectual
    performance and ego depletion: Role of the self in logical reasoning and other
    information processing. *Journal of Personality and Social Psychology,* **85**,
    33–46.

Schwartz, M. B., Chambliss, H. O., Brownell, K. D., Blair, S. N. & Billington, C.
    (2003). Weight bias among health professionals specializing in obesity.
    Obesity Research, **11**, 1033-1039.

Shiffman, S. & Waters, A. J. (2004). Negative affect and smoking lapses: A
    prospective analysis, *Journal of Consulting and Clinical Psychology,* **72**,
    192-201.

Shiffman, S. (2005). Dynamic influences on smoking relapse process, *Journal of
    Personality,* **73**, 1-34.

Shiffman, S., Balabanis, M. H., Gwaltney, C. J., Paty, J. A., Gnys, M., Kassel, J.
    D., Hickcox, M. & Paton, S. M. (2007). Prediction of lapse from associations
    between smoking and situational antecedents assessed by ecological
    momentary assessment. *Drug and Alcohol Dependence*, *91*(2-3), 159-68

Shoda, Y., Mischel, W. & Peake, P. K. (1990). Predicting adolescent cognitive
    and self-regulatory competencies from preschool delay of gratification.
    Developmental Psychology, 26, 978–986.

Spring, B., Pagoto, S., Pingitore, R., Doran, N., Schneider, K. & Hedeker, D.
    (2004). Randomized controlled trial for behavioral smoking and weight

control treatment: effect of concurrent versus sequential intervention. *Journal of Consulting and Clinical Psychology,* **72**, 785-796.

Stucke, T. S. & Baumeister, R. F. (2006). Ego depletion and aggressive behavior: Is the inhibition of aggression a limited resource? *European Journal of Social Psychology,* **36**, 1-13.

Stunkard, A. J. (1959). Obesity and the denial of hunger. *Psychosomatic Medicine,* **21**, 281-289.

Tangney, J. P., Baumeister, R. F. & Boone, A. L. (2004). High self-control predicts good adjustment, less pathology, better grades, and interpersonal success. *Journal of Personality and Social Psychology,* **72**, 271-322.

Teachman, B. A. & Brownell, K. D. (2001). Implicit anit-fat bias among health professionals: Is anyone immune? *International Journal of Obesity Related Metabolic Disorders,* **10**, 1525-31.

Tice, D. M., Baumeister, R. F., Shmueli, D. & Muraven, M. (2007). Restoring the self: Positive affect helps improve self-regulation following ego-depletion, *Journal of Experimental Social Psychology,* **43**, 379-384.

Tiffany, S. T. (1990). A cognitive model of drug urges and drug-use behavior: Role of automatic and nonautomatic processes. *Psychological Review,* **97**, 147-168.

Trope, Y. & Neter, E. (1994). Reconciling competing motives in self-evaluation: The role of self-control in feedback seeking. *Journal of Personality and Social Psychology,* **66**, 1034-1048.

Trope, Y. & Pomerantz, E. M. (1998). Resolving conflicts among self-evaluative motives: Positive experiences as a resource for overcoming defensiveness. *Motivation and Emotion,* **22**, 53-72.

U.S. Department of Agriculture (1998). *Continuing survey of food intakes by individuals,* 1994-1996. U.S. Department of Health and Human Services (2005). National Heart, Lung, and Blood Institute: Your Guide to a Healthy Heart.

U.S. Department of Health and Human Services (2001). *The Surgeon General's call to action to prevent and decrease overweight and obesity.* Rockville, MD: U.S. Department of Health and Human Services, Public Health Service, Office of the Surgeon General.

U.S. Department of Health and Human Services (2004). *The Health Consequences of Smoking: A Report of the Surgeon General.* Atlanta, GA: U.S. Department of Health and Human Services, Centers for Disease Control and Prevention, National Center for Chronic Disease Prevention and Health Promotion, Office on Smoking and Health.

Vohs, K. D. & Heatherton, T. F. (2000). Self-regulatory failure: A resource-depletion approach. *Psychological Science,* **11**, 243-254.

Wegner, D. M., Schneider, D., Carter, S. R. & White, T. L. (1987). Paradoxical effects of thought suppression. *Journal of Personality and Social Psychology,* **53**, 5-13.

Wildes, J. E., Emery, R. E. & Simons, A. D. (2001). The roles of ethnicity and culture in the development of eating disturbance and body dissatisfaction: a meta-analytic review. *Clinical Psychology Review,* **21**, 521–551.

Wills, T. & Shiffman, S. (1985). *Coping and substance use: A conceptual framework.* In S. Shiffman & T. A. Wills (Eds.), Coping and substance use (3-24). New York: Academic.

Williamson, D. F., Madans, J., Anda, R. .F, Kleinman, J. C., Giovino, G. A. & Byers, T. (1991). Smoking Cessation and Severity of Weight Gain in a National Cohort. *New England Journal of Medicine,* **11**, 739–45.

World Health Organization. (2006). *Tobacco free initiative: why is tobacco a public health priority?*

In: Control Theory and Its Applications   ISBN: 978-1-61668-384-9
Editor: Vito G. Massari, pp. 41-53   © 2011 Nova Science Publishers, Inc.

Chapter 2

TRANSACTIONAL PATHWAYS IN THE DEVELOPMENT OF EXTERNALIZING BEHAVIORS IN A SAMPLE OF KINDERGARTEN CHILDREN WITH IMPAIRED SELF-CONTROL

Michael G. Vaughn[1], Brian E. Perron[2], Kevin M. Beaver[3], Matt DeLisi[4]* and Jade Wexler[5]

[1]Saint Louis University,
[2]University of Michigan,
[3]Florida State University,
[4]Iowa State University
[5]University of Texas at Austin

Abstract

Problems relating to self-regulatory skills, interpersonal skills, and learning difficulties place children at increased risk for persistent externalizing behaviors. However, less is known about the antecedents of externalizing behavior in children most at-risk for antisocial behavior over the life-course. The current study used longitudinal data from 1,594 children previously shown to have severe behavioral problems selected from the Early Childhood Longitudinal Survey, Kindergarten Class (ECLS-K) to examine these developmental

* E-mail address: delisi@iastate.edu. Phone: 515-294-8008; Fax: 515-294-2303; (Corresponding author).

pathways. Structural equation modeling showed that learning problems, fine motor problems, and gross motor problems occurring at wave 1 were interrelated and variously predictive of self-control deficits and interpersonal deficits at wave 2. Both self-control and interpersonal deficits at wave 2 significantly predicted externalizing behaviors at wave 4. The findings add to an accumulating knowledge base indicating that externalizing behaviors are importantly related to learning, motor, interpersonal, and self-control deficits.

Keywords: Externalizing behavior, children, impulse control, childhood anti-social behavior, self-control

Introduction

Although forms of self-regulation are basic to an understanding of externalizing behaviors, the interrelationships between self-control variables, early learning abilities, and interpersonal skills are not fully understood. In addition, although there has been extensive research on the externalizing spectrum of behaviors (e.g., Lahey, Moffitt, & Caspi, 2003; Loeber & Farrington, 1998), the transactional nature of the antecedents of externalizing behaviors has not been modeled extensively (Hinshaw, 2002). This constitutes a pressing research need because early development of externalizing behavior is likely to portend a range of maladaptive outcomes such as a heightened risk for life-course antisocial behavior (Campbell, Shaw, & Gilliom, 2000; DeLisi, 2005; Fergusson & Horwood, 1998; Moffitt, 1993). As such, the specification of key mechanisms which cause externalizing behaviors is important to facilitate prevention and treatment interventions.

Based on decades of research from a range of scientific disciplines, it is clear that self-control impairment underpins a substantial portion of the vulnerability to externalizing problem behaviors across the life course (Barkley, 2005; Fishbein, 2000; Gottfredson & Hirschi, 1990). Deficits in self-control have been shown to predict a wide range of externalizing behaviors including criminality and substance abuse (Brower & Price, 2001; DeLisi & Vaughn, 2008; Pratt & Cullen, 2000; Tarter, Kirischi, Mezzich, Cornelius, Pajer, Vanyukov, Gardner, Blackson, & Clark, 2003; Vaughn, Beaver, DeLisi, Perron, & Schelbe, 2009; Vaughn, Beaver, DeLisi, & Wright, 2009; Wills, Ainette, Mendoza, Gibbons, & Brody, 2007; Wills & Dishion, 2004). However, the role of self-control impairment as part of a nexus of interactive risk factors, such as learning skills and interpersonal relations is not clear. On one hand, there is the idea that low self-control produces learning and interpersonal problems. On the other hand, there is a belief that learning, particularly language difficulties and interpersonal deficits contribute to low self-control.

Children who have learning difficulties and impulse-control problems are at greater risk of having interpersonal skills deficits and generally poor relations with others due to a difficult temperament and peer rejection (Dodge & Sherrill, 2007; Lahey & Waldman, 2007; Moffitt, 1993). One consequence of learning problems that may be particularly important to the emergence of self-control and externalizing behavior is language skills (Dionne, Tremblay, Boiven, Laplante, & Perusse, 2003). For instance, using data from twin pairs selected from the Early Childhood Longitudinal Survey, Kindergarten Class (ECLS-K), Beaver, DeLisi, Vaughn, Wright, and Boutwell (2008) found robust cross-sectional and longitudinal relationships between language skills and levels of self-control. Children who scored low on language assessments during wave 1 interviews were more likely to have low levels of self-control at wave 1 and at wave 4. Moreover, they found evidence that covariation in language skills and self-control was largely heritable. At wave 1, 61% of the covariation between language assessments and self-control was due to genetic factors; at wave 4 the heritability estimate was 76%.

Deficits in language skills have been associated with a range of antisocial phenotypes (Moffitt, 1990, 1993; Wilson & Herrnstein, 1985). Learning problems often occur in tandem with a clinical diagnosis of attention-deficit-hyperactivity disorder (ADHD), (Cohen et al., 1998; Willcutt, Pennington, & DeFries, 2000), functional deficits in accurately understanding emotions, and difficulties in social problem solving (Cohen et al., 1998). Learning problems and subsequent poor language abilities also are linked to early-life externalizing behaviors (Gallagher, 1999), physical aggression (Dionne et al., 2003), and delinquency (Davis, Sanger, & Morris-Friehe, 1991). In clinical samples, prevalence estimates have shown that greater than 50% percent of youthful psychiatric patients (Giddan, Milling, & Campbell, 1996) and approximately 80% of antisocial boys have language impairments (Warr-Leeper, Wright, & Mack, 1994). In short, a mélange of problems relating to language skills, self-regulatory skills, interpersonal skills, and learning difficulties place children at increased risk for persistent externalizing behaviors.

Current Focus

The current objective was to model the relationships between motor skills, learning problems, interpersonal skills, and self-control on externalizing behaviors in a large sample of children previously identified with impaired self-control. In a previous study, Vaughn et al. (2009) identified a subgroup of youths (9.8%, n = 1,594) from a kindergarten cohort who demonstrated poor self-control based on parent and teacher assessments over time. Results from a polytomous logistic regression analysis showed that externalizing behaviors, interpersonal skills deficits, and learning problems were important predictors of membership in this severe subgroup. The structural relations between these and other variables; however, was unexamined. As such, the current investigation extended these findings by employing structural equation modeling with the aforementioned variables in a temporally ordered fashion to better understand the antecedents of externalizing behavior in children most at-risk for antisocial behavior over the life-course.

Method

Participants and Procedures

This study uses data from the Early Childhood Longitudinal Survey, Kindergarten Class of 1998-1999 (ECLS-K). The ECLS-K is a nationally-representative cohort of children drawn from approximately 1,000 kindergarten programs and collected by the National Center for Education Statistics (NCES). Six waves of data have been collected thus far with the first wave of data collected in the fall of kindergarten (1998). The second round of interviews was conducted in the spring of 1999 when the children were approaching the end of their first year of formal education. The third and fourth waves of data were collected in the fall of first grade and the spring of first grade, respectively. However, only a small sub-sample of respondents was re-interviewed at wave three. The last two interviews were completed in third grade and in fifth grade. Overall, more than 20,000 children participated in the ECLS-K making it the largest prospective and nationally-representative sample of children. Additional information pertaining to the complex sampling strategy employed and design issues can be found at the NCES (NCES, 1999).

Employing a multi-rater measurement strategy makes the ECLS-K a very rich and appealing data set to examine the manifestations of externalizing behaviors early in life. The data was collected via one-on-one assessment of children, parent, and teacher reports using standardized questionnaires. The mother was the usual parent providing reports. If mothers were not available, the other parent or guardian was interviewed. Response rates for parents were high (> 90%).

Approximately 10% of parent interviews were selected for validation by study field supervisors to assess validity and ensure accuracy. Using both parental and teacher reports has many advantages. One of which is that data collected from teachers can help measure conduct that occurs away from parents (Harris, 1998). As previously described, we used a subsample of 1,594 children (the full analytic sample was 17,212) identified by parents and teachers as having self-control deficits across three waves of data. The majority of children in the analytic sample were male (68.5%) and non-white (53.4%).

Measures

Wave three outcome variable: Externalizing behaviors. Externalizing problem behaviors scale is derived from an adapted version of Gresham and Elliott's (1990) Social Skills Rating Scale (SSRS) included in the ECLS-K. The SSRS is a multi-rater, standardized scale for rating perceptions of subscales of social behavior. This scale is comprised of five items and includes assessments of the frequency of child acting out behaviors such as arguing, fighting, getting angry, and disturbing ongoing activities. Split half reliabilities were good with each administration (α range = .86 - .90).

Wave one variables: Fine and gross motor skills and learning problems. The ECLS-K contains specifically employed assessments of fine and gross motor skills. These were derived from direct one-on-one assessments of children at wave one. The approaches to learning scale based on both teacher and parent reports was used and assesses the extent to which a child's behaviors impacts their learning environment and includes items that rate attentiveness, task persistence, eagerness to learn, learning independence, flexibility, and organization (higher scores indicate greater problems). Due to increased reliability over the parent scale, the teacher rating was employed (α = .89).

Wave two variables: Interpersonal skills and self-control. Several scales were available across three waves to assess behavioral and interpersonal skills. The interpersonal skills scale had five items that rate the child's skill in such tasks as forming friendships, helping other children, and showing sensitivity to the feelings of other people with higher scores indicating greater problems with interpersonal skills. We employed the teacher rating scale (α = .89). Self-control is also derived from the SSRS. Parents and teachers were asked a series of questions, ranging from one (never) to four (very often), pertaining to multiple domains of the child's personality, development, and functioning. Items indicated the child's ability to control behavior by respecting the property rights of others, controlling temper, accepting peer ideas for group activities, and responding appropriately to pressure from peers. The responses to the items were then summed together to form a self-control scale with higher scores indicating lower levels of self control. Reliability was marginally adequate across waves (α range = .60-.66); however, prior researchers analyzing the ECLS-K have used identical measures and research has found the SSRS to be a valid way to measure self-control in children (Beaver et al., 2007, Beaver et al., 2008; Beaver & Wright, 2005).

Control variables: Parental involvement, gender, and race. Three control variables were also included in the analyses to take into account extraneous influences. First, there is some reason to believe that parents are integrally involved in the development of antisocial outcomes (Gottfredson & Hirschi, 1990). As a result, we included a measure of parental involvement. During wave 1 interviews, mothers were asked to indicate how often they engage in nine activities with their child. For example, mothers were asked the frequency with which they read to their child, tell their child stories, and play games with their child, among others. Responses to each of the questions were added together to form the parental involvement scale, where higher values indicated more parental involvement (α = .74). The last two control variables included in the models were gender and race. Both of these variables were dichotomous dummy variables, where gender was coded as 0 = female and 1 = male and race was coded 0 = white and 1 = nonwhite.

Analytic Strategy

The primary analytic strategy for this study was structural equation modeling (SEM). Data were approximately multivariate normal and the default method of estimation was used (i.e., maximum likelihood). Models used variables temporally ordered as follows: Wave 1 (learning problems, fine and gross motor skills), Wave 2 (interpersonal skills deficit and self-control problems), and Wave 3 outcome (externalizing behaviors). We also executed multiple group comparisons to account for the stability in results accounting for variations across gender, race, and parental involvement. The adequacy of model fit was determined using multiple fit indices including the chi-square (χ^2) test, root mean square error of approximation (RMSEA), comparative fit index (CFI), and goodness of fit index (GFI). Given the large sample size, interpretations of coefficients were made based on effect size rather than statistical significance. An effect size of least |.10| was considered clinically significant. All analyses were carried out using LISREL version 8.80.

Results

Baseline Model

Prior to examining the structural relations of study models means, standard deviations and inter-correlations of study variables were computed (see Table 1). Following this procedure, a full structural equation model was estimated, which tested all the hypothesized relationships. The overall model exhibited a good fit evidenced by a non-significant chi-square value ($\chi^2[3] = 5.52$, $p = .142$) and all other fit measures were within a range that indicated excellent model fit (RMSEA = .02, 90% CI = .00 - .05; CFI = 1.00; GFI = 1.00).

The parameter estimates of the baseline model are presented in the path diagram in Figure 1. All statistically significant path associations were in the expected direction. All three exogenonous variables (i.e., learning problems, fine motor problems, gross motor problems) exhibited statistically significant inter-correlations. Fine and gross motor problems at T1 had non-significant paths to self-control and interpersonal deficits at T2. Learning problems at T1 did not have a significant path association with self-control deficits at T2; however, its effect on interpersonal deficits at T2 was significant ($\beta = .30$, $p < .001$). Interpersonal deficits at T2 was also significantly associated with self-control deficits at T2, with a large effect size ($\beta = .69$, $p < .001$). This relationship suggests that the effect of learning problems is fully mediated by interpersonal deficits. Finally, self-control deficits and interpersonal deficits at T2 had significant paths to externalizing behaviors, with self-control deficits having a much larger effect size than interpersonal deficits ($\beta = .19$ versus .06).

Multiple Group Comparisons

Using multiple group comparisons, additional analyses were conducted on the trimmed baseline model to determine the stability of estimates when accounting for gender, race, and parental involvement. All parameters across the groups were allowed to freely vary during the estimation. All models exhibited a good fit with

48   Michael G. Vaughn, Brian E. Perron, Kevin M. Beaver et al.

the data, indicated by non-significant chi-square values and fit indices that were within a range of good fit. All parameter values were similar in magnitude to the baseline model, and no significant path coefficients exhibited a reversal in directions. In each group comparison, no standardized parameter estimate from one group to another changed by a value greater than |.10|.

Table 1. Means, Standard Deviations, and Inter-correlations of Measurement (N = 1594)

Variable	Mean	SD	1.	2.	3.	4.	5.	6.
1. Learning skill problems	2.67	.55	1.00					
2. Fine motor skill problems	4.12	2.14	.39	1.00				
3. Gross motor skill problems	2.07	1.96	.23	.28	1.00			
4. Interpersonal skill deficits	2.62	.49	.30	.12	.07	1.00		
5. Self-control deficits	2.67	.46	.24	.06	.04	.69	1.00	
6. Externalizing behaviors	2.43	.66	.11	.06	.03	.19	.23	1.00

Note: N = 1,594

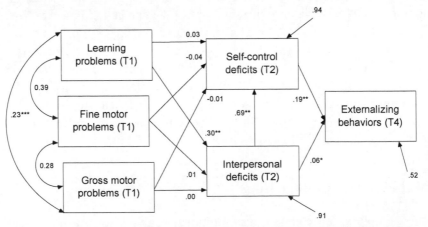

Note: $\chi^2[3] = 5.52$, p = .142; RMSEA = .02, 90% CI = .00 - .05; CFI = 1.00; GFI = 1.00

Figure 1. Path Diagram Showing Structural Associations of Learning and Motor Problems, Interpersonal Deficits, and Externalizing Behaviors among Young Children (N = 1,594).

Conclusion

Antisocial behavior can be relatively stable over long swaths of the life course making the understanding of the early development of externalizing behavioral problems particularly important to understand. An abundance of research has revealed, for instance, that the overwhelming majority of adult criminals have long histories of antisocial behavior that dates back to early childhood (DeLisi, 2005). This finding is of particular importance because it suggests that one way of preventing crime and delinquency is to prevent the emergence of antisocial behaviors. Of course, this entails intimate knowledge about the correlates to childhood misbehavior as well as the different transactional processes that ultimately produce such conduct. The current study sought to shed some light on the interrelationships among learning problems, motor problems, self-control deficits, interpersonal deficits, and externalizing behavioral problems in a large sample of at-risk children.

Analysis of the ECLS-K data revealed that children with lower levels of self-control and with more interpersonal deficits were at risk for evincing externalizing behavioral problems at least one year into the future. Results also identified a strong pathway from early learning difficulties to interpersonal skills deficits. These results not only add to an impressive knowledge base indicating that problems with impulse control, self-regulation, and behavioral inhibition are strong predictors of childhood aggression and antisocial behavior (Séguin, Boulerice, Harden, Tremblay, & Pihl, 1999; Séguin, Nagin, Assaad, & Tremblay, 2004) but also highlight the significance of intervening with early learning difficulties as a means toward preventing externalizing behavior problems in the future. Recent work has shown that infants possess a differential susceptibility to maternal sensitivity as it relates to externalizing problems and that greater maternal sensitivity is associated with reductions in future externalizing behaviors (e.g., Belsky, Fearon, & Bell, 2007; Bradley & Corwyn, 2008). Importantly, our models showed that including parental involvement did not alter the transactional pathways of the final model. Admittedly, the parental involvement measure is not a direct substitute for maternal sensitivity measures yet does indicate the robustness of the identified pathways.

Still, these findings should be interpreted with caution in light of two main limitations. First, although the data are temporally ordered, the time period spans about one and a half years, which is not long enough to assess whether these relationships would remain stable into late childhood, adolescence, and even adulthood. It could be the case that developmental delays in physiological and social maturation may alter the child's current trajectory. Second, while this study is involves a large and representative database, it lacks depth in terms of measurement. For example, there were few neuropsychological tests available in the ECLS-K. Future studies need to address these issues and begin to map out the full breadth of childhood risk factors that contribute to the development of behavioral problems.

References

Barkley, R. A. (2005). *ADHD and the nature of self-control*. New York: The Guilford Press.

Beaver, K. M. & Wright, J. P. (2005). Evaluating the effects of birth complications on low self-control in a sample of twins. *International Journal of Offender Therapy and Comparative Criminology, 49*, 450-471.

Beaver, K. M., Wright, J. P. & DeLisi, M. (2007). Self-control as an executive function: Reformulating Gottfredson and Hirschi's parental socialization thesis. *Criminal Justice and Behavior, 34*, 1345-1361.

Beaver, K. M., DeLisi, M., Vaughn, M. G. Wright, J. P. & Boutwell, B. B. (2008). The relationship between self-control and language: Evidence of a shared etiological pathway. *Criminology, 46*, 201-232.

Belsky, J., Fearon, R. M. P. & Bell, B. (2007). Parenting, attention and externalizing problems: Testing mediation longitudinally, repeatedly and reciprocally. *Journal of Child Psychology and Psychiatry, 48*, 1233-1242.

Bradley, R. H. & Corwyn, R. F. (2008). Infant temperament, parenting, and externalizing behavior in first grade: A test of the differential susceptibility hypothesis. *Journal of Child Psychology and Psychiatry, 49*, 124-131.

Brower, M.C. & Price, B.H. (2001). Neuropsychiatry of frontal lobe dysfunction in violent and criminal behaviour: A critical review. *Journal of Neurology & Neurosurgical Psychiatry, 71*, 720-726.

Campbell, S. B., Shaw, D. S. & Gilliom, M. (2000). Early externalizing behavior problems: Toddlers and preschoolers at risk for later maladjustment. *Development and Psychopathology, 12*, 467-488.

Cohen N. J., Menna, R., Vallance, D. D., Barwick, M. A., Im, N. & Horodezky, N. B. (1998). Language, social cognitive processing, and behavioral characteristics of psychiatrically disturbed children with previously identified and unsuspected language impairments. *Journal of Child Psychology and Psychiatry, 39*, 853-864.

Davis, A. D., Sanger, D. D. & Morris-Friehe, M. (1991). Language skills of delinquent and nondelinquent adolescent males. *Journal of Communication Disorders,* **24**, 251-266.

DeLisi, M. (2005). *Career criminals in society.* Thousand Oaks, CA: Sage.

DeLisi, M. & Vaughn, M. G. (2008). The Gottfredson-Hirschi critiques revisited: Reconciling self-control theory, criminal careers, and career criminals. *International Journal of Offender Therapy and Comparative Criminology,* **52**, 520-537.

Dionne, G. Tremblay, R., Boiven, M., Laplante, D. & Perusse, D. (2003). Physical aggression and expressive vocabulary in 19-month-old twins. *Developmental Psychology,* **39**, 261-273.

Dodge, K. A. & Sherrill, M. R. (2007). The interaction of nature and nurture in antisocial behavior. In D. J. Flannery, A. T. Vazsonyi & I. D. Waldman (Eds.), *The Cambridge handbook of violent behavior and aggression* (Pp. 215-244). New York: Cambridge University Press.

Fergusson, D. M. & Horwood, L. J. (1998). Early conduct problems and later life opportunities. *Journal of Child Psychology and Psychiatry,* **39**, 1097-1108.

Fishbein, D. (2000). Neuropsychological function, drug abuse, and violence: A conceptual framework. *Criminal Justice and Behavior,* **27**, 139-159.

Gallagher, T. M. (1999). Interrelationships among children's language, behavior, and emotional problems. *Topics in Language Disorders,* **19**, 1-15.

Giddan, J. J., Milling, L. & Campbell, N. B. (1996). Unrecognized language and speech deficits in preadolescent psychiatric inpatients. *American Journal of Orthopsychiatry,* **66**, 85-92.

Gottfredson, M. R. & Hirschi, T. (1990). *A general theory of crime.* Stanford, CA: Stanford University Press.

Gresham, F. M. & Elliott, S. N. (1990). *The social skills rating system.* Circle Pines, MN: American Guidance Services.

Harris, J. R. (1998). *The nurture assumption: Why children turn out the way they do.* New York: Free Press.

Hinshaw, S. P. (2002). Process, mechanism, and explanation related to externalizing behavior in developmental psychopathology. *Journal of Abnormal Child Psychology,* **30**, 431-446.

Jöreskog, K. G. & Sörbom, D. *LISREL* **8**: *Structural equation modeling with SIMPLIS command language.* Lincolnwood: Scientific Software International.

Lahey, B. B., Moffitt, T. E. & Caspi, A. (Eds.). (2003). *Causes of conduct disorder and juvenile delinquency.* New York: Guilford Press.

Lahey, B. B. & Waldman, I. D. (2007). Personality dispositions and the development of violence and conduct problems. In D. J. Flannery, A. T. Vazsonyi & I. D. Waldman (Eds.), *The Cambridge handbook of violent behavior and aggression* (Pp. 260-288). New York: Cambridge University Press.

Loeber, R. &. Farrington, D. P. (Eds.). (1998). *Serious and violent juvenile offenders: Risk factors and successful interventions.* Thousand Oaks, CA: Sage.

Moffitt, T. E. (1990). The neuropsychology of juvenile delinquency: A critical review. In N. Morris & M. Tonry (Eds.), *Crime and justice: An annual review of research,* Volume 12 (pp. 99-169). Chicago: University of Chicago Press.

Moffitt, T. E. (1993). Adolescence-limited and life-course persistent antisocial behavior: A developmental taxonomy. *Psychological Review,* **100**, 674-701.

National Center for Education Statistics (NCES). 2001. *ECLS-K base year public-use data files and electronic codebook.* Washington, DC: U.S. Department of Education. Available online at: http://nces.ed.gov/pubs2001/2001029rev_1_4.pdf.

National Center for Education Statistics (NCES). 2002. *Early-childhood longitudinal study-kindergarten class of 1998-99 (ECLS-K), psychometric report for kindergarten through first grade* (Working Paper No. 2002-05). Washington, DC: U.S. Department of Education. Available online at http://nces.ed.gov/pubs2002/200205.pdf.

Pratt, T. C. & Cullen, F. T. (2000). The empirical status of Gottfredson and Hirschi's general theory of crime: A meta-analysis. *Criminology,* **38**, 931-964.

Séguin, J. R., Boulerice, B., Harden, P. W., Tremblay, R. E. & Pihl, R. O. (1999). Executive functions and physical aggression after controlling for attention deficit hyperactivity disorder, general memory, and IQ. *Journal of Child Psychology and Psychiatry,* **40**, 1197-1208.

Séguin, J. R., Nagin, D., Assaad, J. M. & Tremblay, R. E. (2004). Cognitive-neuropsychological function in chronic physical aggression and hyperactivity. *Journal of Abnormal Psychology,* **113**, 603-613.

Tarter, R. E, Kirischi, L., Mezzich, A., Cornelius, J., Pajer, K., Vanyukov, M., Gardner, W., Blackson, T. & Clark, D.B. (2003). Neurobehavior disinhibition in childhood predicts early age onset substance use disorder. *American Journal of Psychiatry,* **160**, 1078-1085.

Vaughn, M. G., Beaver, K. M. & DeLisi, M. & Wright, J. P. (2009). Identifying latent classes of behavioral risk based on early childhood manifestations of self-control. *Youth Violence and Juvenile Justice,* **7**, 16-31.

Vaughn, M. G., Beaver, K. M., DeLisi, M., Perron, B. E. & Schelbe, L. (2009). Gene-environment interplay and the importance of self-control in predicting polydrug use and substance-related problems. *Addictive Behaviors, 34,* 112-116.

Warr-Leeper, G., Wright, N. A. & Mack, A. (1994). Language disabilities of antisocial boys in residential treatment. *Behavioral Disorders, 19,*159-169.

Willcutt, E. G., Pennington, B. F. & DeFries, J. C. (2000). Twin study of the etiology of comorbidity between reading disability and attention-deficit/ hyperactivity disorder. *American Journal of Medical Genetics, 96,* 293-301.

Wills, T. A. & Dishion, T. J. (2004). Temperament and adolescent substance use: A transactional analysis of emerging self-control. *Journal of Clinical Child & Adolescent Psychology, 33,* 69-81.

Wills, T. A., Ainette, M. G., Mendoza, D., Gibbons, F. X. & Brody, G. H. (2007). Self-control, symptomatology, and substance use precursors: Test of a theoretical model in a community sample of 9-year-old children. *Psychology of Addictive Behaviors, 21,* 205-215.

Wilson, J. Q. & Herrnstein, R. J. (1985). *Crime and human nature: The definitive study of the causes of crime.* New York: Simon & Schuster.

In: Control Theory and Its Applications
Editor: Vito G. Massari, pp. 55-81

ISBN: 978-1-61668-384-9
© 2011 Nova Science Publishers, Inc.

Chapter 3

ON THE TRACKING OF A TRAJECTORY AND A CONTROL BY THE FEEDBACK PRINCIPLE*

V.I. Maksimov
Institue of Mathematics and Mechanics,
Ural Branch of the Russian Academy of Sciences,
Ekaterinburg, Russia

Abstract

For dynamical systems described by parabolic differential equations, a problem of the tracking of a trajectory (a problem of etalon motion tracking) and a problem of the tracking of a control (a problem of dynamical input identification) are investigated. Solving algorithms stable with respect to informational noises and computational errors are designed. The algorithms are based on the method of feedback control.

1. Introduction. Problem Statements

In the present paper, we would like to pay attention to the fact that a uniform approach can be used for investigating two rather different problems (namely, a problem of etalon motion tracking and a problem of dynamical input identification). Meaningfully, the essence of the problems under discussion is as follows.

[0]*The work was partly supported by the Russian Foundation for Basic Research (projects no. 10–01–00002 and 07–01–00008), the Program "Mathematical theory of control" of the Presidium of Russian Academy of Sciences, and the Ural-Siberian integration project.

Consider a parabolic equation

$$x_t(t,\eta) - \Delta_L x(t,\eta) = \Phi(x(t,\eta)) + f(t,\eta) \quad \text{in} \quad T \times \Omega = Q, \quad T = [0,\vartheta] \tag{1}$$

with the initial

$$x(0,\eta) = x_0(\eta) \quad \text{in} \quad \Omega$$

and boundary

$$\frac{\partial x(t)}{\partial n}\bigg|_{\Gamma} = u(t) - v(t) \quad \text{in} \quad (0,\vartheta] \times \Gamma \tag{2}$$

conditions. Here $\Omega \subset R^n$ is an open bounded domain with a sufficiently smooth boundary Γ, Δ_L is the Laplace operator, i.e.,

$$\Delta_L x(\eta) = \sum_{i=1}^{n} \frac{\partial^2 x(\eta)}{\partial \eta_i^2},$$

$\eta = (\eta_1, \ldots, \eta_n)$, $x_0(\eta) \in L_2(\Omega)$, $\Phi(\cdot)$ is a Lipschits function, $u(\cdot) \in L_2(T; L_2(\Gamma))$ is a control, $v(\cdot) \in L_2(T; L_2(\Gamma))$ is a disturbance. On the time interval T, a uniform partition

$$\Delta = \{\tau_i\}_{i=0}^{m}, \quad \tau_i = \tau_{i-1} + \delta,$$

$\tau_0 = 0$, $\tau_m = \vartheta$, is fixed. A solution of equation (1) $x(\cdot) = x(\cdot; 0, x_0, u(\cdot), v(\cdot))$ depends on a varying in time control $u(\cdot)$ and an unknown disturbance $v(\cdot)$. The function $x(\cdot)$ is also unknown. At the moments $\tau_i \in \Delta$, phase states $x(\tau_i)$ are inaccurately measured. Results of measurements, i.e., elements $\xi_i^h \in H = L_2(\Omega)$, $i \in [1 : m-1]$, satisfy the inequalities

$$|\xi_i^h - x(\tau_i)|_H \leq h. \tag{3}$$

Here $h \in (0,1)$ is a value of informational accuracy. In what follows, we assume for simplicity that the initial state x_0 is known.

A problem of etalon motion tracking. It is assumed that $v = v(t) \equiv 0$, $t \in T$, in the right-hand part of boundary conditions (2). Let a number $\varepsilon > 0$ be given. There is an etalon motion, which is described also by an equation of the form (1) with $v \equiv 0$ and $u = u^*(t)$. Both the function $u^*(\cdot)$ and the solution $g(\cdot)$ of the etalon equation are unknown. It is only known that $u^*(t) \in D_*$ for almost all $t \in T$, where $D_* \subset L_2(\Gamma)$ is a given set. At the moments $\tau_i \in \Delta$, along with $x(\tau_i)$, states $g(\tau_i)$ are inaccurately measured. It is required to design an algorithm for forming a feedback control $u = u(t) \in P$, $t \in T$, such that the solution of equation (1) with boundary condition (2) remains at some " ε-neighborhood" of the etalon motion for all $t \in T$.

A problem of dynamical inversion. (A problem of dynamical input identification). Let in the right-hand part of boundary conditions (2) the control is equal to zero, i.e., $u = u(t) = 0, t \in T$. Let $Q \subset L_2(T)$ is a bounded closed and convex set. It is required to design a dynamical algorithm that allows us to reconstruct an unknown input (a disturbance) $v = v(\cdot)$ $(v(t) \in Q$ for almost all $t \in T)$ in "real time" mode.

For solving problems of all two types described above, one can use a uniform approach based on the method of auxiliary position-controlled models. It should be noted that the laws for choosing model controls are based on different modifications of the principle of extremal shift. The method of extremal shift is one of the most effective methods for investigating feedback control problems. This method was suggested by N.N. Krasovskii [1]. Then it has been widely applied to investigating game control problems as well.

Let us illustrate the essence of the method by the following example. Consider a problem of motion tracking for a system Σ of the form

$$\dot{z}(t) = f(t, z(t)) + B(u(t) - v(t)), \quad t \in T,$$

where $x \in R^n$ is a phase state, $u, v \in R^m$, $z(0) = z_0$, B is an $n \times m$-dimensional matrix, a function f is Lipschitz with respect to the set of variables, $v(t) \in Q$ is a disturbance, $u(t) \in P$ is a control, $P, Q \subset R^m$ are bounded closed sets. It is required to construct a law for choosing a control

$$u = u(t, z, y)$$

such that the corresponding trajectory of the system Σ is close (in the uniform metric) to the trajectory of the system

$$\dot{y}(t) = f(t, y(t)), \quad t \in T, \quad y(0) = z_0,$$

i. e., the value

$$I(x, y) = \sup_{t \in T} |z(t; u(\cdot), v(\cdot)) - y(t)|_{R^n}$$

is small.

Let $Q \subset P$ and a partition $\Delta = \{\tau_i\}_{i=0}^m$, $\tau_0 = 0$, $\tau_m = \vartheta$, with a diameter $\delta = \vartheta/m$ be fixed. Let us assume that the trajectory $z(t)$ is measured at the moments τ_i with an error h. Results of measurements, vectors $\tilde{\xi}_i^h \in R^n$, satisfy the inequalities

$$|\tilde{\xi}_i^h - z(\tau_i)|_{R^n} \le h.$$

The essence of the method of extremal shift with respect to the problem in question consists in the choice of the control u in the following form:

$$u(\tau_i, \tilde{\xi}_i^h, y(\tau_i)) = \arg \min\{(\tilde{\xi}_i^h - y(\tau_i), Bu)_{R^n} : u \in P\}.$$

The symbols $|\cdot|_{R^n}$ and $(\cdot, \cdot)_{R^n}$ stand for the norm and the scalar product in R^n, respectively.

If

$$u(t) = u(\tau_i, \tilde{\xi}_i^h, y(\tau_i)), \quad t \in [\tau_i, \tau_{i+1}),$$

then, as follows from results of [1], for any $\varepsilon > 0$ there exist $h_1 > 0$ and $\delta_1 > 0$ such that if $h \in (0, h_1)$, $\delta \in (0, \delta_1)$, then the inequality

$$\sup_{t \in T} |z(t; u(\cdot), v(\cdot)) - y(t)|_{R^n} \le \varepsilon$$

is valid for any disturbance $v(t) \in Q$. Also, for other classes of systems, analogous problems were discussed from the position of the method of extremal shift in [2, 3].

Both problems described above are solved with the use of the same scheme.

The scheme of algorithm for solving the problem of dynamical input identification is shown in the Fig. 1.

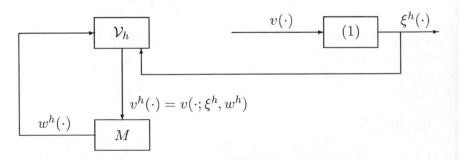

Figure 1.

In the beginning, an auxiliary system M (called a model) is introduced. The model has an input $v^h(\cdot)$ and an output $w^h(\cdot)$. The process of synchronous feedback control of equation (1) and M is organized on the interval T. This process is decomposed into $(m-1)$ identical steps. At the i-th step carried out during the time interval $\delta_i = [\tau_i, \tau_{i+1})$, the following actions are fulfilled. First, at the time moment τ_i, according to some chosen rule $\mathcal{V}_h$, the element

$$v_i^h = \mathcal{V}_h(\tau_i, \xi_i^h, w^h(\tau_i))$$

is calculated. Then (till the moment τ_{i+1}) the control $v^h(t) = v_i^h$, $\tau_i \le t < \tau_{i+1}$, is fed onto the input of equation (1). The values ξ_{i+1}^h and $w_{i+1}^h = w^h(\tau_{i+1})$ are treated as algorithm's output at the i-th step.

An analogous scheme is applicable to solving the problem of etalon motion tracking (see Fig. 2).

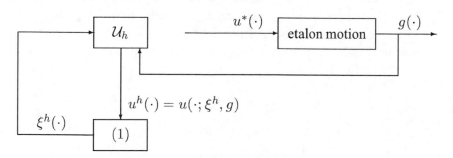

Figure 2.

In this case, etalon motion plays the role of the model, i.e. the model (etalon motion) has an input $u^*(\cdot)$ and an output $g(\cdot)$. The process of synchronous feedback control of equation (1) and etalon equation is organized on the interval T. This process is also decomposed into $(m-1)$ identical steps. At the i-th step carried out during the time interval $\delta_i = [\tau_i, \tau_{i+1})$, the following actions are fulfilled. First, at the time moment τ_i, according to some chosen rule $\mathcal{U}_h$, the element

$$u_i^h = \mathcal{U}_h(\tau_i, \xi_i^h, g(\tau_i))$$

is calculated. Then (till the moment τ_{i+1}) the control $u^h(t) = u_i^h$, $\tau_i \le t < \tau_{i+1}$, is fed onto the input of equation (1). The values ξ_{i+1}^h and $g_{i+1} = g(\tau_{i+1})$ are treated as algorithm's output at the i-th step.

As is known, a solution of a parabolic equation of the form (1) with a boundary control of the form (2) can be defined in different ways. It is possible to consider a classical solution [4] (its existence requires fulfillment of rather severe constraints on a domain Ω, an initial state x_0, a control u, and a disturbance v) or a generalized solution, used, as a rule, in control theory. The latter, in turn, can be introduced differently. Most often, to define a generalized solution, the approach developed, for example, in [5, 6] (sometimes it is called the "functional analytic" approach) as well as the semigroup approach [7–11] (rather actively popularized in recent years and originating from [12]) are used. In the present paper, we use both these approaches to the definition of a solution.

2. Solving the Problem of Etalon Motion Tracking

In this section, we consider the case when $\Phi(z) = x$, i.e., system (1) is linear. Before passing to the description of an algorithm for solving the problem formulated above, we give the rigorous definition of a solution of equation (1) with boundary conditions (2). Introduce the Neumann mapping $\mathcal{N}$:

$$\mathcal{N}v = z \Longleftrightarrow \begin{cases} \Delta_L z - z = 0 \quad \text{in} \quad \Omega, \\[2mm] \dfrac{\partial z}{\partial n}\bigg|_\Gamma = v. \end{cases} \tag{4}$$

In other words, $\mathcal{N}v$ is a generalized solution of elliptic equation (4), i.e., a function with the following properties:

$$\mathcal{N}v \in H = L_2(\Omega), \quad \frac{\partial(\mathcal{N}v)}{\partial n} \in L_2(\Gamma),$$

$$\int_\Omega (\mathcal{N}v)(\eta)\{\Delta_L\psi(\eta) + \psi(\eta)\}\,d\eta = \int_\Gamma \psi(\sigma)\frac{\partial(\mathcal{N}v)(\sigma)}{\partial n}\,d\sigma \quad \forall\psi \in H_2(\Omega).$$

Under a solution of equation (1), (2) corresponding to a control $u(\cdot) \in L_\infty(T;U)$ and a disturbance $v(\cdot) \in L_\infty(T;U)$, we understand, according to [8–10], a unique function $x(\cdot) = x(\cdot; 0, x_0, u(\cdot), v(\cdot)) \in C(T;H)$ of the form

$$x(t; t_0, x_0, u(\cdot), v(\cdot)) = S(t)x_0 + A\int_0^t S(t-\tau)\mathcal{N}(u(\tau) - v(\tau))\,d\tau, \quad t \in T.$$

$$\tag{5}$$

Here and below, we assume that

$$U = L_2(\Gamma), \quad H = L_2(\Omega),$$

$$Ah = \Delta_L h - h, \qquad h \in \mathcal{D}(A) = \left\{ z \in H_2(\Omega) : \frac{\partial z}{\partial n}\Big|_\Gamma = 0 \right\}$$

is the infinitesimal generator of analytical contracting semigroup of linear continuous operators $\{S(t); t \geq 0\}$ on H. In what follows, we, as is customary [8–10], identify A with its isomorphic extension $A : H \to \mathcal{D}^*(A)$. As is known, all eigenvalues of the operator A belong to the interval $(0, +\infty)$. This fact provides the bounded invertibility of the operator A, i.e.,

$$A^{-1} \in \mathcal{L}(H; H)$$

[9, p. 291; 10, p. 54]. From [10, p. 54] it follows also that $\mathcal{N}$ is a continuous mapping of the space $L_2(\Gamma)$ into the space

$$H^{3/2}(\Omega) \subset H^{3/2-2\varepsilon}(\Omega) = \mathcal{D}(A^{3/4-\varepsilon}) \quad \forall \varepsilon > 0.$$

In addition [11, p. 292], the inclusions

$$\mathcal{N} \in \mathcal{L}(H^s(\Gamma); H^{s+3/2}(\Omega)) \quad \forall s \in R, \quad A^{-l} \in \mathcal{L}(H^{2\alpha}(\Omega); H^{2\alpha+2l}(\Omega))$$

are valid for any nonnegative integers l and any $\alpha \in R^+ = [0, +\infty)$.

The symbols $H^\alpha(\Omega)$ and $H^s(\Gamma)$ stand for the standard Sobolev spaces.

Thus, an etalon motion is given. This motion is described by the equation

$$g_t(t, \eta) - \Delta_L g(t, \eta) + g(t, \eta) = f(t, \eta) \quad \text{in} \quad T \times \Omega, \tag{6}$$

$$g(0, \eta) = x_0(\eta) \quad \text{in} \quad \Omega$$

with the Neumann boundary condition

$$\frac{\partial g(t)}{\partial n}\Big|_\Gamma = u^*(t) \quad \text{in} \quad (0, \vartheta] \times \Gamma. \tag{7}$$

Here $u^*(t) \in D_* \subset U$, D_* is a given bounded and closed set, the function $u^*(\cdot)$ (Lebesque measurable) is unknown.

The rule for choosing a control $u = u^h(\cdot)$ (for every $h \in (0, 1)$) in system (1) is identified with a pair $S_h = (\Delta_h, \mathcal{U}_h)$, where

$$\Delta_h = \{\tau_{h,i}\}_{i=0}^{m_h} \tag{8}$$

is a partition of the interval T into half-intervals $[\tau_{h,i}, \tau_{h,i+1})$, $\tau_{h,i+1} = \tau_{h,i} + \delta$, $\delta = \delta(h)$, $\tau_{h,0} = 0$, $\tau_{h,m_h} = \vartheta$, $\mathcal{U}_h$ is a mapping assigning to every triple $(\tau_i, \xi_i^h, g(\tau_i))$, $i \in [0 : m_h - 1]$, $\tau_i = \tau_{h,i}$ (called a position) an element

$$u_i^h = \mathcal{U}_h(\tau_i, \xi_i^h, g(\tau_i)) \in D_*. \tag{9}$$

Here $\tau_i = \tau_{h,i}$, Ξ_T is the set of measurements, i.e., the set of all piece-wise constant functions $\xi(\cdot) : T \to H$,

$$\Xi(x(\cdot), h)$$

is the set of all h-accurate results of measurements, i.e., the set of all functions $\xi^h(\cdot) \in \Xi_T$ satisfying (3).

Under an "ε-neighborhood" of a point $z \in H$, we understand the set of elements $y \in H$ such that

$$|A^{-1}(z - y)|_H \le \varepsilon.$$

Introduce the following functional Λ^0:

$$\Lambda^0(x(\cdot), g(\cdot)) = \sup_{t \in T} |A^{-1}(g(t) - x(t))|_H.$$

Now we describe an algorithm for choosing a control $u(\cdot) = u^h(\cdot)$ in system (1) with the boundary condition

$$\left. \frac{\partial x(t)}{\partial n} \right|_\Gamma = u^h(t) \quad \text{in} \quad (0, \vartheta] \times \Gamma \tag{10}$$

providing retention of the solution $x^h(\cdot) = x(\cdot; 0, x_0, u^h(\cdot))$ (of equation (1) with boundary conditions (10)) in some "ε-neighborhood" of $g(\cdot)$ for all $t \in T$.

Let the mapping $\mathcal{U}_h$ from (9) be given by the formula

$$\mathcal{U}_h(\tau_i, \xi_i^h, g(\tau_i)) = \{u_i^h : (s_i^*, \mathcal{N}u^h)_H \le \inf\{(s_i^*, \mathcal{N}v)_H : v \in D_*\} + h\},$$

where

$$s_i^* = A^{-1}(g(\tau_i) - \xi_i^h).$$

In this case

$$u^h(t) = u_i^h, \quad t \in [\tau_i, \tau_{i+1}), \quad \tau_i = \tau_{h,i}, \quad i \in [0 : m_h - 1]. \tag{11}$$

Theorem 1 *Whatever $\varepsilon > 0$ and $\xi^h(\cdot) \in \Xi(x(\cdot), h)$ may be, one can find a number $h_* > 0$ such that for all $h \in (0, h_*)$ the rule for choosing a control $u = u^h$ (in system (1) with boundary conditions (10)) of the form (8), (9), (11) provides fulfillment of the inequality*

$$\Lambda^0(x^h(\cdot), g(\cdot)) \leq \varepsilon.$$

Thus, the rule S_h solves the problem of tracking the etalon motion $g(\cdot)$. Before proving the theorem, we give two auxiliary statements.

Lemma 1 *There exists a number $d_* = d_*(x_0)$ such that the estimate*

$$|x(t; 0, x_0, u(\cdot))|_H \leq d_*(x_0)\{1 + |u(\cdot)|_{L_\infty(T;U)}\} \tag{12}$$

holds for any $t \in T$ and $u(\cdot) \in L_\infty(T; U)$.

The lemma follows from inequality (3.22) [13, p. 233]:

$$|AS(t)\mathcal{N}v|_H \leq C(t)|v|_U, \quad t > 0, \quad v \in U, \quad C(\cdot) \in L_2(T; R), \tag{13}$$

the contractability of the semigroup $\{S(t); t \geq 0\}$, and the Gronwall inequality [14, p. 219].

Let a bounded set $Q_1 \subset L_2(T; U)$ be fixed. By the symbol $W(x_0, Q_1)$ we denote a bundle of solutions of equation (1) with boundary condition (2) for $v(t) = 0$ corresponding to all controls $u(\cdot) \in Q_1$, i.e.,

$$W(x_0, Q_1) = \{x(\cdot; 0, x_0, u(\cdot)) : u(\cdot) \in Q_1\} \subset C(T; H).$$

Introduce the notation

$$A^{-1}W(x_0, Q_1) = \{\tilde{x}(\cdot) : \tilde{x}(t) = A^{-1}x(t) \ \forall t \in T, x(\cdot) \in W(x_0, Q_1)\} \subset C(T; H).$$

Lemma 2 *The set $A^{-1}W(x_0, Q_1)$ is uniformly bounded and equicontinuous in the space $C(T; H)$.*

Proof of theorem 1. Consider the variation of the value

$$\Lambda(t) \equiv \Lambda(t, x^h(\cdot), g(\cdot)) = |A^{-1}(x^h(t) - g(t))|_H^2.$$

It is easily seen that the following inequality is valid:

$$\Lambda(\tau_{i+1}) = |A^{-1}\{S(\delta)(x^h(\tau_i) - g(\tau_i)) \tag{14}$$

$$+ A \int_0^\delta S(\delta - \tau)\mathcal{N}(u^h(\tau_i + \tau) - u^*(\tau_i + \tau))\,d\tau|_H^2 \le \sum_{j=1}^4 J_{ji},$$

where

$$J_{1i} = |s_i|_H^2, \qquad s_i = A^{-1}S(\delta)(x^h(\tau_i) - g(\tau_i)),$$

$$J_{2i} = 2\left(s_i, \int_0^\delta S(\delta - \tau)\mathcal{N}(u^h(\tau_i + \tau) - u^*(\tau_i + \tau))\,d\tau\right)_H,$$

$$J_{3i} = \left|\int_0^\delta S(\delta - \tau)\mathcal{N}(u^h(\tau_i + \tau) - u^*(\tau_i + \tau))\,d\tau\right|_H^2.$$

Since the semigroup $\{S(t); t \ge 0\}$ is contracting, the operator A^{-1} commutates with $S(\delta)$, the inequality

$$J_{1i} \le |A^{-1}(g(\tau_i) - x^h(\tau_i))|_H^2 \tag{15}$$

holds. In virtue of the inclusions

$$A^{-1} \in \mathcal{L}(H; H), \quad \mathcal{N} \in \mathcal{L}(L_2(\Gamma); H),$$

the inequality

$$J_{3i} \le c_0\delta^2 \tag{16}$$

is also valid. Hereinafter, the symbol c_j denotes a constant not depending on $\xi^h(\cdot)$ and $u^h(\cdot)$, which can ce explicitly written. Note that [9, p. 314]

$$A \int_0^t S(t - s)x\,ds = S(t)x - x \qquad \forall\, x \in H.$$

Therefore, taking into account the contractability of the semigroup $S(t)$, we obtain

$$|A^{-1}\{S(t)x - x\}|_H \le t|x|_H. \tag{17}$$

Then, using (3), lemma 1, and (17), we get the following estimates:

$$|s_i - \tilde{s}_i|_H = |A^{-1}\{S(\delta)(x^h(\tau_i) - g(\tau_i)) - (x^h(\tau_i) - g(\tau_i))\}|_H \qquad (18)$$

$$\le \delta|g(\tau_i) - x^h(\tau_i)|_H \le c_1\delta,$$

$$|\tilde{s}_i - s_i^*|_H \le c_2 h,$$

where

$$\tilde{s}_i = A^{-1}(x^h(\tau_i) - g(\tau_i)).$$

Consequently, taking into account the self-conjugacy of the semigroup $\{S(t); t \ge 0\}$ and inequality (18), we derive

$$J_{2i} \le 2\left(\tilde{s}_i, \int_0^\delta S(\delta - \tau)\mathcal{N}(u^h(\tau_i + \tau) - u^*(\tau_i + \tau))\, d\tau\right)_H + c_3\delta^2 \qquad (19)$$

$$= 2\int_0^\delta (A^{-1}S(\delta - \tau)(x^h(\tau_i) - g(\tau_i)), \mathcal{N}(u^h(\tau_i + \tau) - u^*(\tau_i + \tau)))_H \, d\tau + c_3\delta^2$$

$$\le 2\int_0^\delta (A^{-1}(x^h(\tau_i) - g(\tau_i)), \mathcal{N}(u^h(\tau_i + \tau) - u^*(\tau_i + \tau)))_H \, d\tau + c_4\delta^2$$

$$\le 2\int_0^\delta (s_i^*, \mathcal{N}(u^h(\tau_i + \tau) - u^*(\tau_i + \tau)))_H \, d\tau + c_5\delta(\delta + h).$$

In virtue of the rule for choosing a control $u^h(\cdot)$ (see (11)), from (19) we conclude that for all $h \in (0, h^*)$

$$J_{2i} \le c_6\delta(\delta + h) \quad i \in [0 : m - 1]. \qquad (20)$$

Combining estimates (14)–(16) and (3), we derive

$$\Lambda(\tau_{i+1}) \le \Lambda(\tau_i) + c_7\delta\{h + \delta)\} \le c_7\nu(h, \delta)\delta,$$

where

$$\nu(h, \delta) = h + \delta.$$

Thus,

$$\Lambda(\tau_i) \le c_7 \vartheta (h + \delta), \quad i \in [1 : m]. \tag{21}$$

Since the constant c_7 can be explicitly written, the assertions of the theorem follow from estimates (21) and lemma 2. The theorem is proved.

Remark 1. Introduce the conjugated operator $\mathcal{N}^* \in \mathcal{L}(H; L_2(\Gamma))$:

$$(\mathcal{N}v, y)_H = (v, \mathcal{N}^* y)_{L_2(\Gamma)} \quad \forall v \in L_2(\Gamma), \quad y \in H.$$

Then, by the Green formula, we have

$$\mathcal{N}^* y = \left(\Delta_L^{-1} y \right)\big|_\Gamma \quad \forall y \in H.$$

Let us describe the algorithm for solving a problem. Before the initial moment the value h and the partition $\Delta = \Delta_h$ with diameter $\delta = \delta(h)$ are fixed. The work of the algorithm starting at time $t = 0$ is decomposed into $m_h - 1$ steps. At the i-th step carried out during the time interval $\delta_i = [\tau_i, \tau_{i+1})$, $\tau_i = \tau_{h,i}$, the following actions take place. First, at time moment τ_i element u_i^h is calculated by the formulas (9), (11). Then the controls (11) are fed onto the input of the equation (1). After that, we transform the state $x(\tau_i)$ of the equation (1) into $x(\tau_{i+1})$. The procedure stops at time ϑ.

3.   Solving the Problem of Dynamical Inversion

Let us turn to the problem of dynamical inversion, which belongs to the class of inverse problems and, in a more general context, to the class of ill-posed problems (see [15–16]). Similar problems in an *a posteriori* statement have been investigated by many scientists. In [2], a method of dynamical reconstruction of an input to a finite-dimensional dynamical system affine in disturbance was suggested. The case when a set $P \subset R^m$ of "instantaneous" restrictions on $u(t)$ was fixed (i.e., $u(t) \in P$ for $t \in T$) was considered. This method was further developed in a number of works. The method is based on the ideas of the theory of positional control (see [1]) in the combination with the methods of the smoothing functional and discrepancy (see [15, 16]) known in the theory of ill-posed problems. Thus, consider equation (1) with the boundary condition

$$\frac{\partial x(t)}{\partial n}\bigg|_\Gamma = v(t) \quad \text{in} \quad (0, \vartheta] \times \Gamma, \tag{22}$$

where $v(t) \in Q$ for a.a. $t \in T$, $Q \subset L_2(T)$ is a bounded and closed set. For the sake of simplicity, we omit the minus in the right-hand part of (2).

Consider the mapping

$$t \to p(t; \cdot, \cdot, \cdot) : H \times L_2(T; U) \times C(T; H) \to C(T; H),$$

$$p(t; x_0, u(\cdot), z(\cdot)) \;=\; S(t)x_0 + A \int_0^t S(t - \tau) \mathcal{N} u(\tau)\, d\tau$$

$$+ \int_0^t S(t - \tau) \Phi_1(z(\tau))\, d\tau, \quad t \in T.$$

Here

$$Ah = \Delta_L h - h, \qquad h \in \mathcal{D}(A) = \left\{ z \in H_2(\Omega) : \left.\frac{\partial z}{\partial n}\right|_\Gamma = 0 \right\},$$

and A is the infinitesimal generator of analytical contracting semigroup of linear continuous operators $\{S(t); t \geq 0\}$ on H, $\Phi_1(x) = \Phi(x) + x$.

According to [7], a solution of problem (1), (22) corresponding to a disturbance $v(\cdot) \in L_\infty(T; U)$ is a unique function $x(\cdot) = x(\cdot; 0, x_0, v(\cdot)) \in C(T; H)$ satisfying the equality

$$x(t) = p(t; x_0, v(\cdot), x(\cdot)) \quad \forall t \in T.$$

Let $V(x(\cdot))$ be the set of all disturbances compatible with the output $x(\cdot)$. In other words,

$$V(x(\cdot)) \;=\; \left\{ v(\cdot) \in V_T : x(t) - S(t)x_0 - \int_0^t S(t - \tau)\Phi_1(x(\tau))\, d\tau = \right.$$

$$= \; A \int_0^t S(t - \tau)\mathcal{N} v(\tau)\, d\tau \quad \forall t \in T \Big\},$$

where

$$V_T = \{ v(\cdot) \in L_2(T; U) : v(t) \in Q \quad \text{for almost all} \quad t \in T \}.$$

It is easily seen that this set is convex, bounded, and closed in $L_2(T;U)$. Therefore, it contains a unique element $v_*(\cdot) = v_*(\cdot; x(\cdot))$ of minimal $L_2(T;U)$-norm. Let us describe an algorithm for reconstructing

$$v_*(\cdot) = v_*(\cdot; x(\cdot)) = \arg\min\{|v(\cdot)|_{L_2(T;U)} : v(\cdot) \in V(x(\cdot))\}.$$

Let $\varphi_x(\cdot)$ be the continuity modulus of the function $t \to \Phi_1(x(t)) \in H$ in T, i.e.,

$$\varphi_x(\delta) = \sup\{|\Phi_1(x(t_1)) - \Phi_1(x(t_2))|_H : t_1, t_2 \in T, |t_1 - t_2| < \delta\}.$$

Let a family $\{\Delta_h\}$ of partitions of the interval T of the form (8) as well as a function $\alpha(h) : R^+ \to R^+ = \{r \in R : r > 0\}$ satisfying the conditions

$$\alpha(h) \to 0, \quad \delta(h) \to 0,$$
$$(h + \varphi_x(\delta(h)))\alpha^{-1}(h) \to 0 \quad \text{as } h \to 0+ \tag{23}$$

be chosen. Introduce an auxiliary system described by the linear equation (a model M)

$$w_t^h(t, \eta) - \Delta_L w^h(t, \eta) + w^h(t, \eta) = f(t, \eta) + \Phi_1(\xi_i^h(\eta)) \tag{24}$$

$$\text{in } [\tau_{h,i}, \tau_{h,i+1}) \times \Omega, \ i = 0, 1, \ldots, m_h - 1,$$
$$w^h(0, \eta) = x_0(\eta) \quad \text{in} \quad \Omega$$

with the Neumann boundary condition

$$\left.\frac{\partial w^h(t)}{\partial n}\right|_\Gamma = v^h(t) \quad \text{in} \quad (0, \vartheta] \times \Gamma.$$

A solution of equation (24) corresponding to control $v^h(\cdot) \in L_\infty(T;U)$ is a function

$$w^h(\cdot) = w(\cdot; t_0, x_0, w_0^h, \xi^h(\cdot), v^h(\cdot)) \in C(T; H),$$

defined by the equality [8–10]

$$w^h(t) = S(t) + A \int_0^t S(t-\tau)\mathcal{N}v^h(\tau)\,d\tau + \int_0^t S(t-\tau)\Phi_1(\xi^h(\tau))\,d\tau, \quad t \in T.$$

where

$$\xi^h(\tau) = \xi_i^h \quad \text{for} \quad \tau \in [\tau_i, \tau_{i+1}), \quad \tau_i = \tau_{h,i}.$$

The rule for choosing a control $v = v^h(\cdot)$ (at every $h \in (0,1)$) in model (24), as in the problem of etalon motion tracking, is identified with a pair

$$S_h = (\Delta_h, \mathcal{V}_h),$$

where $\Delta_h = \{\tau_{h,i}\}_{i=0}^{m_h}$ is a partition of the interval T into half-intervals $[\tau_{h,i}, \tau_{h,i+1})$, $\tau_{h,i+1} = \tau_{h,i} + \delta$, $\delta = \delta(h)$, $\tau_{h,0} = 0$, $\tau_{h,m_h} = \vartheta$, $\mathcal{V}_h$ is a mapping assigning to every triple $(\tau_i, \xi_i^h, w^h(\tau_i))$, $i \in [0 : m_h - 1]$, $\tau_i = \tau_{h,i}$, an element

$$v_i^h = \mathcal{V}_h(\tau_i, \xi_i^h, w^h(\tau_i)) \in Q.$$

Let the rule for choosing a control $v^h(\cdot)$ in equation (24) $S_h = (\Delta_h, \mathcal{V}_h)$ is given by the formula

$$\mathcal{V}_h(\tau_i, \xi_i^h, w^h(\tau_i)) =$$

$$= [v_i : 2(s_i^*, \mathcal{N}v_i)_H + \alpha(h)|v_i|_U^2 \le \inf\{2(s_i^*, \mathcal{N}v)_H + \alpha(h)|v|_U^2 : v \in Q\} + h],$$

where

$$s_i^* = A^{-1}(w^h(\tau_i) - \xi_i^h).$$

In this case

$$v^h(t) = v^h, \quad t \in [\tau_i, \tau_{i+1}), \quad i \in [0 : m_h - 1].$$

The following theorem is valid.

Theorem 2 *Let condition (23) be fulfilled. Then the convergence*

$$v^h(\cdot) \to v_*(\cdot) \quad \text{in} \quad L_2(T; U) \quad \text{as} \quad h \to 0+$$

holds.

The theorem follows from Lemmas 2.1 and 2.6 [3], and from the next lemma.

Lemma 3 *The following inequalities:*

$$|v^h(\cdot)|_{L_2(T;U)} \le k_1(h)|v_*(\cdot; x(\cdot))|_{L_2(T;U)} + k_2(h),$$

$$\Lambda^0(x(\cdot), w^h(\cdot)) = \sup_{t \in T} \Lambda(t, x(\cdot), w^h(\cdot)) \le k_3(h)$$

are true.

Here

$$\Lambda(t, x(\cdot), w^h(\cdot)) = |A^{-1}(w^h(t) - x(t))|_H^2,$$

functions $k_1(\cdot), k_2(\cdot), k_3(\cdot)$ are such that

$$k_1(h) \to 1, \quad k_2(h) \to +0, \quad k_3(h) \to +0 \quad \text{as} \quad h \to +0.$$

Proof. Consider the value

$$\varepsilon_h(t) = \Lambda(t, x(\cdot), w^h(\cdot)) + \alpha(h) \int_0^t \{|v^h(\tau)|_U^2 - |v_*(\tau)|_U^2\} \, d\tau.$$

It is easily seen that the following inequality is valid:

$$\varepsilon_{i+1} = \left| A^{-1} \{ S(\delta)(w^h(\tau_i) - x(\tau_i)) + A \int_0^\delta S(\delta - \tau) \mathcal{N}(v^h(\tau_i + \tau) - v_*(\tau_i + \tau)) \, d\tau \right.$$

$$+ \int_0^\delta S(\delta - \tau)[\Phi_1(\xi_i^h) - \Phi_1(x(\tau_i + \tau))] \, d\tau \} \bigg|_H^2 + \alpha(h) \int_0^{\tau_{i+1}} \{|v^h(s)|_U^2 - |v_*(s)|_U^2\} \, ds$$

$$\leq \sum_{j=1}^4 J_{ji} + \alpha(h) \int_0^{\tau_{i+1}} \{|v^h(s)|_U^2 - |v_*(s)|_U^2\} \, ds, \qquad (25)$$

where

$$\delta = \tau_{i+1} - \tau_i, \quad \tau_i = \tau_{h,i}, \quad J_{1i} = |s_i|_H^2, \quad s_i = A^{-1}S(\delta)(w^h(\tau_i) - x(\tau_i)),$$

$$J_{2i} = 2\left(s_i, \int_0^\delta S(\delta - \tau) \mathcal{N}(v^h(\tau_i + \tau) - v_*(\tau_i + \tau)) \, d\tau\right)_H,$$

$$J_{3i} = 2\left(s_i, A^{-1} \int_0^\delta S(\delta - \tau)\{\Phi_1(\xi_i^h) - \Phi_1(x(\tau_i + \tau))\} \, d\tau\right)_H,$$

$$J_{4i} = 2\left\{ \left| \int_0^\delta S(\delta - \tau) \mathcal{N}(v^h(\tau_i + \tau) - v_*(\tau_i + \tau)) \, d\tau \right|_H^2 \right.$$

$$+ \left| A^{-1} \int_0^\delta S(\delta - \tau)\{\Phi_1(\xi_i^h) - \Phi_1(x(\tau_i + \tau))\} \, d\tau \right|_H^2 \Biggr\}.$$

The inequalities

$$J_{1i} \le |A^{-1}(w^h(\tau_i) - x(\tau_i))|_H^2, \tag{26}$$

$$|\Phi_1(\xi_i^h) - \Phi_1(x(\tau_i + \tau))|_H \le L(h + \varphi_x(\delta)), \quad \tau \in [0, \delta],$$

hold since the semigroup $\{S(t); t \ge 0\}$ is contracting, the operator A^{-1} commutates with $S(\delta)$, and the function $\Phi_1(\cdot)$ is Lipschitz. Then, by virtue of Lemma 1, we have

$$|\Phi_1(\xi_i^h)|_H \le |\Phi_1(0)|_H + L|\xi_i^h|_H \le |\Phi_1(0)|_H + L\{|x(\tau_i)|_H + h\} \le c_1$$

$$\text{for} \quad t \in \delta_i = [\tau_i, \tau_{i+1}), \ \tau_i = \tau_{h,i}, \quad i \in [0 : m - 1],$$

$$|v^h(t)|_U \le \sup\{|u|_U : u \in P\} \quad \text{for} \quad t \in T.$$

In this case,

$$|w^h(t)|_H \le c_2$$

is true. Hereinafter, the symbols c_j denote constants not depending on $\xi^h(\cdot)$ and $v^h(\cdot)$. Such constants can be explicitly written. Consider the value J_{3i}. We have

$$J_{3i} \le 2\delta a_1 L |(w^h(\tau_i) - x(\tau_i))|_H (h + \varphi_x(\delta)) \le c_3 \delta(h + \varphi_x(\delta)), \tag{27}$$

where

$$a_1 = |A^{-1}|_{\mathcal{L}(H;H)}^2.$$

Moreover, using the inclusions $A^{-1} \in \mathcal{L}(H; H), \mathcal{N} \in \mathcal{L}(L_2(\Gamma); H)$, we obtain the inequality

$$J_{4i} \le c_4 \delta^2. \tag{28}$$

Then (see (18)) we have

$$|s_i - \tilde{s}_i|_H = |A^{-1}\{S(\delta)(w^h(\tau_i) - x(\tau_i)) - (w^h(\tau_i) - x(\tau_i))\}|_H$$

$$\leq \delta |w^h(\tau_i) - x(\tau_i)|_H \leq c_5\delta,$$

$$|\tilde{s}_i - s_i^*|_H \leq c_6 h, \quad \tilde{s}_i = A^{-1}(w^h(\tau_i) - x(\tau_i)). \tag{29}$$

Consequently, taking into account self-conjugacy of the semigroup $\{S(t); t \geq 0\}$ and inequality (29), we derive

$$J_{2i} \leq 2\left(\tilde{s}_i, \int_0^\delta S(\delta - \tau)\mathcal{N}(v^h(\tau_i + \tau) - v_*(\tau_i + \tau))\, d\tau\right)_H + c_6\delta^2$$

$$= 2\int_0^\delta \left(A^{-1}S(\delta - \tau)(w^h(\tau_i) - x(\tau_i)), \mathcal{N}(v^h(\tau_i + \tau) - v_*(\tau_i + \tau))\right)_H d\tau + c_6\delta^2$$

$$\leq 2\int_0^\delta \left(A^{-1}(w^h(\tau_i) - x(\tau_i)), \mathcal{N}(v^h(\tau_i + \tau) - v_*(\tau_i + \tau))\right)_H d\tau + c_7\delta^2$$

$$\leq 2\int_0^\delta \left(s_i^*, \mathcal{N}(v^h(\tau_i + \tau) - v_*(\tau_i + \tau))\right)_H d\tau + c_8\delta(\delta + h). \tag{30}$$

Thus we obtain from (30) the inequality

$$J_{2i} + \alpha(h) \int_0^{\tau_{i+1}} \{|v^h(s)|_U^2 - |v_*(s)|_U^2\}\, ds$$

$$\leq \alpha(h) \int_0^{\tau_i} \{|v^h(s)|_U^2 - |v_*(s)|_U^2\}\, ds + 2\int_0^\delta \left(s_i^*, \mathcal{N}(v^h(\tau_i + s)\right.$$

$$\left. - v_*(\tau_i + s))\right)_H ds + \alpha(h)\int_{\tau_i}^{\tau_{i+1}} \{|v^h(s)|_U^2 - |v_*(s)|_U^2\}\, ds + c_8\delta(\delta + h)$$

$$\leq c_9\delta(\delta+h) + \alpha(h)\int_0^{\tau_i}\{|v^h(s)|_U^2 - |v_*(s)|_U^2\}\,ds, \quad i \in [0:m-1]. \quad (31)$$

Combining estimates (25)–(28) and (31), we derive

$$\varepsilon_{i+1} \leq \varepsilon_i + c_8\delta\{h + \varphi_x(\delta) + \delta\} \leq c_9\nu(h,\delta),$$

where

$$\nu(h,\delta) = h + \varphi_x(\delta) + \delta.$$

We see that the first inequality of the lemma is valid, if

$$k_1(h) = 1, \quad k_2(h) = c_9\nu(h,\delta(h))\alpha^{-1}(h).$$

In addition, by virtue of the relation between parameters (23), $k_2(h) \to 0$ as $h \to 0$. The second inequality follows from Lemma 2. The theorem is proved.

We develop another algorithm for solving the problem in question. Hereinafter, we assume that $V = H^1(\Omega)$, the symbol $y(\sigma)$ stands for the trace of a function $y(\eta) \in H^1(\Omega)$, the symbol ∇x denotes the gradient of x, i.e.,

$$\nabla x = \{\partial x/\partial\eta_1,\ldots,\partial x/\partial\eta_n\},$$

and the symbol $\langle\cdot,\cdot\rangle$ means the duality relation between V and V^*. Let the results of measurements satisfies the inequality

$$|\xi_i^h - x(\tau_i)|_V \leq h.$$

In this case, a solution of problem (1), (22) corresponding to a disturbance $v(\cdot) \in L_\infty(T;U)$ is, according to [5, 6], a unique function $x(\cdot) = x(\cdot;0,x_0,v(\cdot)) \in W(T;V)$ such that

$$d(x(t),y)/dt + \int_\Omega \nabla x(t,\eta)\nabla y(\eta)\,d\eta + \int_\Gamma v(t,\sigma)y(\sigma)\,d\sigma \quad (32)$$

$$= \int_\Omega \Phi(x(t,\eta))y(\eta)\,d\eta \quad \forall y \in H^1(\Omega) \quad \text{for almost all } t \in T.$$

Here,

$$W(T;V) = \{w(\cdot) \in L_2(T;V) : w_t(\cdot) \in L_2(T;V^*)\}.$$

The existence and uniqueness of a solution with the properties above follows from, for example, Statement 1 [17, p. 129].

Such a definition of a solution implies that the set $U(x(\cdot))$ of all inputs $v(\cdot) \in L_2(T; U)$ compatible with $x(\cdot)$ is the set of all $v(\cdot)$ from $L_2(T; U)$ that satisfy variational equality (32) for $x(\cdot) = x(\cdot; 0, x_0, v(\cdot))$. It is easily seen that this set is convex and closed in the space $L_2(T; U)$. Therefore, it contains a unique element $v_*(\cdot) = v_*(\cdot; x(\cdot))$ of minimal $L_2(T; U)$-norm. Now, we consider the case when

$$x(\cdot) \in W^{1,2}(T; V) = \{x(\cdot) \in L_2(T; V) : x_t(\cdot) \in L_2(T; V)\}.$$

Let us describe the algorithm of reconstruction of $v_*(\cdot) = v_*(\cdot; x(\cdot))$. We fix a family $\{\Delta_h\}$ of partitions of the interval T of the form (8); a family of functions $\{\omega_j\}_{j=1}^{\infty} \subset V$ such that $|\omega_j|_V = 1$ and linear combinations of ω_j form a set that is everywhere dense in V;

$$0 < \alpha_j < 1, \quad \alpha_j \to 0, \quad \sum_{j=1}^{\infty} \alpha_j^2 < \infty;$$

and function $\alpha(h) : R^+ \to R^+$ satisfying the following properties:

$$\alpha(h) \to 0, \quad \delta(h) \to 0,$$
$$\delta(h)\alpha^{-1}(h) \to 0, \quad h^2\delta^{-1}(h)\alpha^{-1}(h) \to 0 \text{ as } h \to +0. \tag{33}$$

Introduce the operator $B_1 \in \mathcal{L}(U; V^*)$:

$$\langle B_1 u, y \rangle = \int_{\Gamma} u(\sigma) y(\sigma) \, d\sigma \quad \forall y \in V.$$

Let $|\cdot|_\alpha$ and $\langle \cdot, \cdot \rangle_\alpha$ be the α-norm and the corresponding scalar product in V^* [18, 19] of the form

$$|y|_\alpha = \left(\sum_{j=1}^{\infty} \alpha_j^2 \langle y, \omega_j \rangle^2 \right)^{1/2}, \quad \langle x, y \rangle_\alpha = \sum_{j=1}^{\infty} \alpha_j^2 \langle y, \omega_j \rangle \langle x, \omega_j \rangle \quad \forall x, y \in V^*.$$

Some properties of the α-norm are given, for example, in [3]. In particular, the following inequality is valid:

$$|B_1 v|_\alpha \leq K_1 |v|_U \quad \forall v \in U,$$

where the constant K_1 does not depend on v and α_j.

As a model M, we take in the space V^* an auxiliary controlled system of the form

$$\dot{w}^h(t) = B_1 v^h(t), \quad t \in T,$$
$$w^h(0) = w_0^h = 0 \in V^*.$$

Let a mapping be given by the formula

$$v_i^h = \mathcal{V}_h(\tau_i, \xi_0^h, \ldots, \xi_{i-1}^h, \xi_i^h, w^h(\tau_i)) = \arg\min\{2\langle s_i^0, B_1 v\rangle_\alpha + \alpha|v|_U^2 : v \in Q\}, \tag{34}$$

$$\alpha = \alpha(h), \quad s_i^0 = w^h(\tau_i) - \xi_i^{(1)}.$$

Here the element $\xi_i^{(1)} \in V^*$ is found according to the rule

$$\langle \xi_i^{(1)}, \omega_j \rangle = (\xi_i^h - \xi_0^h, \omega_j)_H + \sum_{j=0}^{i-1} \delta \int_\Omega \{\nabla \xi_j^h(\eta)\nabla\omega_j(\eta) + \Phi(\xi_i^h(\eta))\omega_j(\eta)\}\, d\eta,$$

$$\forall j \in [1:\infty).$$

Note that

$$\langle s_i^0, B_1 v\rangle_\alpha = \sum_{j=1}^\infty \alpha_j^2 \langle s_i^0, \omega_j\rangle\langle B_1 v, \omega_j\rangle = \sum_{j=1}^\infty \alpha_j^2 \langle s_i^0, \omega_j\rangle(v, \omega_j|_\Gamma)_{L_2(\Gamma)} = (v, \varrho_i)_U,$$

where

$$\varrho_i = \sum_{j=1}^\infty \alpha_j^2 \langle s_i^0, \omega_j\rangle\omega_j|_\Gamma.$$

Thus, to find the element v_i^h, there is no need to know $\xi_i^{(1)} \in V^*$, it is enough to be able to calculate elements $\varrho_i \in U$.

In this case

$$v^h(t) = v^h, \quad t \in [\tau_i, \tau_{i+1}), \quad i \in [0:m_h - 1].$$

Theorem 3 *Let condition (33) be fulfilled. Then the convergence*

$$v^h(\cdot) \to v_*(\cdot) \quad in \quad L_2(T;U) \quad as \quad h \to 0+$$

holds.

The theorem follows from Lemmas 2.1 and 2.6 [3], and from the next lemma.

Lemma 4 *One can find constants d_j ($j \in [1:4]$) and $h_* > 0$, not depending on h, δ, α, such that if $\delta\alpha^{-1} \in (0, d_*^{-1})$, $d_* = 4K_1^2$, then for all $h \in (0, h_*)$ the following inequalities are true:*

$$|v^h(\cdot)|^2_{L_2(T;U)} \le d_1 \frac{1 + \delta\alpha^{-1}}{1 - d_*\delta\alpha^{-1}} |v_*(\cdot)|^2_{L_2(T;U)} + d_2 \frac{h + \delta\alpha^{-1} + h^2\delta^{-1}\alpha^{-1}}{1 - d_*\delta\alpha^{-1}},$$
(35)

$$i \in [1 : m_h - 1],$$

$$R(v^h(\cdot), v_*(\cdot), \delta) = \max_{i \in [0:m_h]} \left| \int_0^{\tau_i} B_1\{v^h(\tau) - v_*(\tau)\} \, d\tau \right|_\alpha^2$$

$$\le d_3(h^2\delta^{-1} + \delta) + d_4\delta \int_0^\vartheta \{|v^h(\tau)|^2_U + (1 + \alpha\delta^{-1})|v_*(\tau)|^2_U\} \, d\tau. \tag{36}$$

Proof. Let us estimate the variation of the value

$$\lambda_i^h = \lambda^h(\tau_i) = \left| \int_0^{\tau_i} B_1\{v^h(\tau) - v_*(\tau)\} \, d\tau \right|_\alpha^2, \quad i \in [1 : m_h - 1].$$

We obtain

$$\lambda_{i+1}^h = |w_{i+1}^h - x_{i+1}^{(1)}|_\alpha^2 = \lambda_i^h + I_{1,i}^h + I_{2,i}^h, \tag{37}$$

where

$$I_{1,i}^h = 2\langle w_i^h - x_i^{(1)}, \int_{\tau_i}^{\tau_{i+1}} B_1\{v^h(\tau) - v_*(\tau)\} \, d\tau \rangle_\alpha,$$

$$I_{2,i}^h = \left| \int_{\tau_i}^{\tau_{i+1}} B_1\{v^h(\tau) - v_*(\tau)\} \, d\tau \right|_\alpha^2, \quad w_i^h = w^h(\tau_i).$$

The element $x_i^{(1)} \in V^*$ is determined as follows:

$$\langle x_i^{(1)}, \omega_j \rangle = (x(\tau_i) - x_0, \omega_j)_H + \int_0^{\tau_i}\int_\Omega \{\nabla x(t, \eta)\nabla\omega_j(\eta)$$

$$+ \Phi(x(t, \eta))\omega_j(\eta)\} \, d\eta \, dt \quad \forall j \in [1 : \infty).$$

Note that

$$\langle x_i^{(1)}, \omega_j \rangle = \int_0^{\tau_i} \langle B_1 v_*(\tau), \omega_j \rangle \, d\tau.$$

Thus,

$$\left\langle \int_{\tau_i}^{\tau_{i+1}} B_1 v_*(\tau) d\tau, \omega_j \right\rangle = \langle x_{i+1}^{(1)} - x_i^{(1)}, \omega_j \rangle, \tag{38}$$

$$\left| \int_{\tau_i}^{\tau_{i+1}} B_1 v_*(\tau) d\tau - x_{i+1}^{(1)} + x_i^{(1)} \right|_\alpha = 0.$$

Let

$$\xi^h(t) = \xi_i^h \quad \text{for} \quad t \in [\tau_i, \tau_{i+1}).$$

By virtue of

$$|\xi_i^h - x(\tau_i)|_V \le h,$$

we have

$$|x(t) - \xi^h(t)|_V \le |x(\tau_i) - \xi_i^h|_V + \int_{\tau_i}^{\tau_{i+1}} |\dot{x}(t)|_V \, dt \le h + \int_{\tau_i}^{\tau_{i+1}} |\dot{x}(t)|_V \, dt, \quad t \in \delta_i.$$

In this case,

$$\|x(\cdot) - \xi^h(\cdot)\|_{L_2(T;V)} \le \left(\delta \sum_{i=0}^{m_h-1} \left(h + \int_{\tau_i}^{\tau_{i+1}} |\dot{x}(t)|_V \, dt \right)^2 \right)^{1/2}$$

$$\le \left(2\delta \sum_{i=0}^{m_h-1} \left(h^2 + \delta \int_{\tau_i}^{\tau_{i+1}} |\dot{x}(t)|_V^2 \right) \right)^{1/2} \le c_0(h + \delta).$$

Hereinafter, the symbol c_j stands for constants, which can be explicitly written. It follows that

$$|x_i^{(1)} - \xi_i^{(1)}|_\alpha \le \nu = c_1(h + \delta). \tag{39}$$

Consequently,

$$I_{1,i}^h \le I_{3,i}^h + \beta_{h,i}, \quad \beta_{h,i} = \nu(I_{2,i})^{1/2} \le \nu^2 + I_{2,i}, \tag{40}$$

where

$$I_{3,i}^h = 2\Big\langle w_i^h - \xi_i^{(1)}, \int_{\tau_i}^{\tau_{i+1}} B_1\{v^h(\tau) - u_*(\tau)\}\, d\tau \Big\rangle_\alpha.$$

In addition,

$$\Big| \int_{\tau_{i-1}}^{\tau_i} B_1 u_*(\tau)\, d\tau - \xi_i^{(1)} + \xi_{i-1}^{(1)} \Big|_\alpha \le 2c_1(h+\delta).$$

The last inequality follows from (38), (39). Introduce the function

$$L_i(v) = 2\Big\langle w_i^h - \xi_i^{(1)}, \int_{\tau_i}^{\tau_{i+1}} B_1\{v - v_*(\tau)\}d\tau \Big\rangle_\alpha + \alpha \int_{\tau_i}^{\tau_{i+1}} \{|v|_U^2 - |v_*(\tau)|_U^2\}\, d\tau.$$

By the definition of the control v_i^h (see (34)), in this case, we have

$$L_i(v_i^h) \le h\delta. \tag{41}$$

From (37), (40), and (41) we derive

$$\varepsilon_{i+1}^h \equiv \lambda_{i+1}^h + \alpha \int_0^{\tau_{i+1}} \{|v^h(\tau)|_U^2 - |v_*(\tau)|_U^2\}\, d\tau \le \varepsilon_i + \nu^2 + 2I_{2,i} + h\delta. \tag{42}$$

Then,

$$I_{2,i}^h \;\le\; (\tau_{i+1} - \tau_i) \int_{\tau_i}^{\tau_{i+1}} |B_1(v_i^h - v_*(\tau))|_\alpha^2\, d\tau$$

$$\le 2(\tau_{i+1} - \tau_i) K_1^2 \int_{\tau_i}^{\tau_{i+1}} \{|v_i^h|_U^2 + |v_*(\tau)|_U^2\}\, d\tau. \tag{43}$$

Summing over i the right-hand and left-hand parts of (42) and taking into account (37) and (43), we obtain

$$\varepsilon_{j+1}^h \le \varepsilon_0^h + (\nu^2 + h\delta)j + 4\delta K_1^2 \Big\{ \int_0^{\tau_{j+1}} \{|v^h(\tau)|_U^2 + |v_*(\tau)|_U^2\}\, d\tau \Big\}. \tag{44}$$

Observing that

$$\lambda_j^h \geq 0 \quad \forall j \in [0:m], \quad \varepsilon_0^h = 0,$$

we derive from (44)

$$\alpha \int_0^\vartheta \{|v^h(\tau)|_U^2 - |v_*(\tau)|_U^2\}d\tau \leq (\nu^2 + h\delta)(m-1) + 4K_1^2\delta\left\{\int_0^\vartheta |v^h(\tau)|_U^2\,d\tau\right.$$

$$\left. + \int_0^\vartheta |v_*(\tau)|_U^2\,d\tau\right\}.$$

Hence, by virtue of (39), we conclude that

$$\int_0^\vartheta |v^h(\tau)|_U^2\,d\tau \leq \left(\{c_1^2(h+\delta)^2 + h\delta\}\vartheta\delta^{-1}\right.$$

$$+(\alpha + 4K_1^2\delta)\int_0^\vartheta |v_*(\tau)|_U^2\,d\tau\right)(\alpha - 4K_1^2\delta)^{-1}$$

$$= \left(2c_1^2\vartheta(h^2\delta^{-1}\alpha^{-1} + \delta\alpha^{-1}) + h\vartheta\right.$$

$$+(1 + 4K_1\delta\alpha^{-1})\int_0^\vartheta |v_*(\tau)|_U^2\,d\tau\right)(1 - 4K_1^2\delta\alpha^{-1})^{-1}.$$

Thus, inequality (35) is proved. It is easily seen that one can assume $d_1 = \max\{1, 4K_1^2\}$, $d_2 = \vartheta\max\{1, 2c_1^2\}$. From (44) we obtain the inequality

$$\left|\int_0^{\tau_i} B_1\{v^h(\tau) - v_*(\tau)\}d\tau\right|_{\alpha,h} \leq c_1^2\vartheta(h+\delta)^2\delta^{-1} + h\vartheta$$

$$+ 4K_1^2\delta\int_0^\vartheta |v^h(\tau)|_U^2 d\tau + (\alpha + 4K_1^2\delta)\int_0^\vartheta |v_*(\tau)|^2 d\tau. \tag{45}$$

Inequality (36) is a consequence of relation (45) with a corresponding choice of constants. The lemma is proved.

References

[1] Krasovskii, N.N.; Subbotin, A.I. *Positional Differential Games;* Nauka: Moscow, RU, 1974; 451 p. [in Russian].

[2] Osipov, Yu.S.; Kryazhimskii, A.V. *Inverse Problems for Ordinary Differential Equations: Dynamical Solutions;* Gordon and Breach: London, UK, 1995; 826 p.

[3] Maksimov, V.I. *Dynamical Inverse Problems of Distributed Systems;* Utrecht-Boston: VSP, 2002; 268 p.

[4] Mikhailov, V.P. *Partial Differential Equations;* Nauka: Moscow, RU, 1983; 435 p. [in Russian].

[5] Ladyzhenskaya, O.A. *Boundary-Value Problems of Mathematical Physics;* Nauka: Moscow, RU, 1973; 631 p. [in Russian].

[6] Lions, J.-L. Contrôle optimal de systèmes gouvernés par des équations aux dérivées partielles; Dunod Gauthier-Villars: Paris, FR, 1968; 325 p.

[7] Lasiecka I.; Triggiani, R. *Appl. Math. and Optim.* 1991, vol.23, No.2, 109–154.

[8] Lasiecka, I. *Appl. Math. and Optim.* 1978, vol.4, No.4, 301–328.

[9] Lasiecka, I. *Appl. Math. and Optim.* 1980, vol. 6, No.4, 287–334.

[10] Lasiecka I.; Triggiani, R. *Lecture Notes in Control and Information Sciences.* 1991, vol.164.

[11] Tröltzsch, F. *ZAMM (Z. Angew. Math. Mech.).* 1992, vol.72, No.7, 291–301.

[12] Fattorini, H. *J. Different. Equations.* 1968, vol.5, 72–105; 1969, vol.6, 50–70.

[13] Barbu, V. *SIAM J. Control Optim.* 1980, vol. 18, No.2, 227–243.

[14] Varga, J. *Optimal Control in Differential and Functional Equations;* Mir: Moscow, RU, 1977; 671 p. [in Russian].

[15] Ivanov, V.K.; Vasin, V.V.; Tanana, V.P. *Theory of Linear Ill-Posed Problems and its Applications;* Utrecht-Boston: VSP, 1996; 365 p.

[16] Tikhonov, A.N.; Arsenin, V.Ya. *Methods for Solving Ill-Posed Problems;* Nauka: Moscow, RU, 1978; 268 p. [in Russian].

[17] Barbu, V. *Optimal Control of Variational Inequalities.* Pitman: London, UK, 1984; 485 p.

[18] Osipov, Yu.S.; Okhezin, S.P. *Dokl. Akad. Nauk SSSR.* 1976, vol. 226, No.6, 1267. [in Russian].

[19] Osipov, Yu.S.; Kryazhimskii, A.V.; Maksimov, V.I. In book: *Problems of Dynamical Regularization for Systems with Distributed Parameters;* Izdatel'stvo IMM UrO RAN: Sverdlovsk, RU, 1991; pp. 5–31. [in Russian].

In: Control Theory and Its Applications   ISBN 978-1-61668-384-9
Editor: Vito G. Massari, pp. 83-144   © 2011 Nova Science Publishers, Inc.

Chapter 4

PERTURBATION METHOD IN THE THEORY OF PONTRYAGIN MAXIMUM PRINCIPLE FOR OPTIMAL CONTROL OF DIVERGENT SEMILINEAR HYPERBOLIC EQUATIONS WITH POINTWISE STATE CONSTRAINTS[*]

V.S. Gavrilov[†] and M.I. Sumin[‡]
Mathematical Department, Nizhnii Novgorod State University,
23, Gagarin street, Nizhnii Novgorod 603950, Russia

Abstract

The chapter deals with the parametric nonlinear suboptimal control of divergent hyperbolic equations with pointwise state constraints. Based on the ideology of the perturbation method, we study the suboptimal control problem depending on the infinite–dimensional parameter which is additively contained in the pointwise inequality constraints.

---

[*]This work was supported by the Russian Foundation for Basic Research (project no. 07-01-00495), by the analytical targeted program "Development of the Scientific Potential of Higher Schools (2009 2010)" of the Ministry for Education and Science (project no. 2.1.1/3927), and by Federal target program "Scientific and research and educational manpower of innovative Russia (2009-2013)" (project no. NK-13P-13).

[†]E-mail address: vladimir.s.gavrilov@gmail.com
[‡]E-mail address: m.sumin@mm.unn.ru; msumin@sinn.ru

The minimizing sequence of usual admissible controls is the basic element of our theory. Accordingly, as a minimizing sequence, we use the so-called minimizing approximate solution in the sense of Warga, but not classical minimizing sequence. We study the following issues: finding of suboptimality, regularity, and normality conditions; perturbation theory problems; differential properties of value functions (S–functions); optimal values stability (sensitivity). Detailed examination of these issues as applied to optimal control problems with pointwise state constraints is important both from the standpoint of the general theory of distributed optimization and from the standpoint of constructing various efficient numerical algorithms for solving optimization problems for such systems. For calculation of functionals' first variations, we use the so-called two–parametrical needle variation of admissible controls. Besides, we discuss an approximation of the source problem by problems with a finite number of functional constraints.

Keywords: optimal control, parametric problem, minimizing sequence, perturbation method, necessary and sufficient suboptimality conditions, sensitivity, Pontryagin maximum principle, semilinear divergent hyperbolic equations, linear divergent hyperbolic equations with measures in input data, pointwise state constraints, two–parametrical needle variation

Introduction

This chapter is devoted to continuation of research of the authors in the field of parametric optimal control problems for partial differential equations. It deals with the parametric nonlinear suboptimal control of divergent hyperbolic equations with pointwise state constraints. Based on the ideology of the perturbation method, we study the suboptimal control problem depending on infinite–dimensional parameter which is additively contained in the pointwise inequality constraints.

Optimal control problems with pointwise state constraints have been of interest for forty years. Research in this area began naturally with optimal control problems for systems of ordinary differential equations. Active studying optimization problems for distributed parameters systems with pointwise state constraints began more than two decades ago, and interest in such problems has remained stable since that time (see, e.g., works [1] – [12] and bibliographies of these works).

However, the majority of publications on optimal control of distributed systems with pointwise state constraints are devoted to finding only necessary optimality conditions, in particular, those similar to Pontryagin's maximum principle (see, e.g., works [1] – [12]). Other classic optimization problems related to the specified class of systems have received little attention in the literature. Among them are problems of finding of suboptimality, regularity, and normality conditions; perturbation theory problems; differential properties of value functions (S–functions); optimal values stability (sensitivity). Detailed examination of these issues as applied to optimal control problems with pointwise state constraints is important both from the standpoint of the general theory of distributed optimization and from the standpoint of constructing various efficient numerical algorithms for solving optimization problems for such systems. These issues were considered, e.g., in works [13] – [16] for elliptic partial differential equations, and in works [17]–[19] for nonlinear hyperbolic Goursat–Darboux system with pointwise state constraints. At the same time, we do not know of any works in which these theory issues are considered in the case of divergent hyperbolic partial differential equations. This fact is explained by the difficulty of studying such problems for hyperbolic divergent partial differential equations. The difficulty is that, even in relatively simple cases, solutions to hyperbolic divergent equations may not possess regularity properties, which are typical for solutions to elliptic and parabolic equations (see, e.g., [20]). In particular, these typical regularity properties were essentially used in the above–mentioned works [1] – [10].

What was said above is our motivation for studying an optimization problem for a nonlinear divergent hyperbolic equations with pointwise state constraints containing an additive functional parameter. The minimizing sequence of usual admissible controls is the basic element of our theory. Accordingly, as a minimizing sequence, we use the so-called minimizing approximate solution (m.a.s.) in the sense of Warga [21], but not classical minimizing sequence.

Here we use the so-called two parametric (multipoint) variation of admissible controls. This variation was suggested in works [22] – [25]. The basic idea is that first variations are computed using only the intrinsic properties of solutions as representatives of the classes of functions to which they belong. As applied to problems for divergent equations, these are the classes of functions possessing generalized Sobolev derivatives. For functions from Sobolev spaces, their intrinsic property underlying the needle variation modification under consideration is a classical property (see, e.g., [26]) according to which the differential structure of such functions in directions where generalized derivatives exist is better than their structure with respect to the collection of arguments. Need of such modification of a needle variation is explained by the following two reasons. Firstly, this is explained by a non-stability of classic Lebesgue points (see, e.g., [25]) of functions belonging to Sobolev classes with respect to perturbations of these functions in metrics of Sobolev classes. Secondly, this is explained by "worse" properties of solutions to divergent hyperbolic equations than properties of solutions to elliptic and parabolic equations. This fact leads to essential difficulties, if we use classic needle variation for a calculation of first variations of optimization problem's functionals. For simplest optimal control problem for divergent hyperbolic equation, sufficiently detailed discussion of these circumstances and other needed explanations connected with the variation method used here can be found in [25].

The needle variation modification in question is based on a two-parameter method varying controls. Accordingly, the first variation is understood as a repeated limit. It can be effectively used to derive necessary optimality conditions in various optimal control problems for both linear and nonlinear (quasilinear) partial differential equations. The one-dimensional variation parameter is split into two parameters, of which the first is dictated by the directions in the space of independent variables in which the solutions to differential adjoint equations have generalized derivatives. The first of two passages to the limit in two-parameter variation is with respect to this variation parameter. Simultaneously, the two-parameter modification of needle variation naturally leads to a modification of the classical concept of a Lebesgue point (see, e.g., [27]), which, in contrast to its classical analogue, is stable in a natural sense with respect to a perturbation of the function in the Sobolev class containing the solution to the adjoint problem.

In the chapter, the method of research of the considered (sub)optimal control problem with pointwise state constraints is based on the method of the work [28]. This method can be split into three main parts.

1) We approximate the source problem with pointwise state constraints by problems, where each of approximating problems is "equivalent" to a problem with finite number of functional constraints. Here the source problem with pointwise state constraints is interpreted as a problem with infinite number of functional inequality constraints. This approximation allows to use advantages of finite–dimensional non–smooth analysis by comparison with infinite–dimensional non–smooth analysis. In partic-

ular, one of such essential advantages from the standpoint of obtaining sensitivity results is that, if a function f of n variables is lower semi-continuous, and has a bounded subdifferential in the sense Clarke [29] or Mordukhovich [30] at a point $\bar{x}$, then f is Lipschitz in a neighborhood of the point $\bar{x}$ (see [29, Proposition 2.9.7], [30], [31], [32, Corrolary 8.5])*.

2) Using two–parametrical needle variation of admissible controls [25], in each approximating problem, we obtain (as in optimal control problems with finite number of functional inequality constraints) "approximating" maximum principle. Its formulation involves adjoint functions for each of finite number of functional inequality constraints. These adjoint functions are solutions to ordinary adjoint hyperbolic partial differential equations in a divergence form.

3) Going to the limit as a number of functional inequality constraints tends to infinity, we get a resulting maximum principle for the source problem with pointwise state constraints. For this, we "fuse" family of adjoint equations (each of adjoint equations corresponds to some functional inequality constraint) into one adjoint equation. The "fused" adjoint equation corresponds to the source state constraint, and involves a Radon measure in the right–hand side part.

The chapter consists of an introduction and six basic sections. In Section 1, we give the statement of the parametric nonlinear suboptimal control problem for divergent hyperbolic partial differential equation with pointwise state inequality constraint. Section 2 contains formulations of all basic results of the work. Section 3 provides the necessary auxiliary results. In the section we study issues needed for obtaining main results. These issues are connected with a theory of linear hyperbolic divergent equations involving a Radon measure in the right–hand side part. Such equations appears (in the form of adjoint equations of a maximum principle) in the proof of a Pontryagin maximum principle (or a generalizations of it) for optimal control problems with pointwise state constraints. We study the following questions: existence, uniqueness and stability of solutions to such equations; special integral representations of solutions to such equations; and a stability of solutions linear hyperbolic divergent equations with respect to initial conditions on a hyperplane $t = \tau$ (but not top or bottom of a cylinder $Q_T \equiv \Omega \times (0, T)$), and with respect to a position of a

---

*Generally speaking, this fact is not valid in the infinite–dimensional case.

hyperplane $t = \tau$. Let us note that we do not know any analogous results concerning with linear hyperbolic divergent equations involving a Radon measure in the right–hand side part. Besides, in this section we pay attention to the problem of a calculation of first variations of functionals. For effective solution of this problem under natural assumptions of input data of an optimization problem with pointwise state constraints, we use a modification of a classic multi-point needle variation. Section 4 deals with the proof of suboptimality necessary conditions, in the form generalizing a classical Pontryagin maximum principle, in the considered problem. Interpreting the source problem with pointwise state constraints as a problem with infinite number of functional constraints, in Section 5 we discuss an approximation of the source problem by problems with finite number of functional constraints. At last, in Section 6, we prove normality and regularity conditions, results connected with differential properties of the value function, and sufficient suboptimality conditions.

1.   Problem Statement

We begin with some notation. Suppose $U \subset R^m$ is a compact set, $V \subset R$ is a segment, $A,\ T > 0$ are constants, $\Omega \subset R^n$ is a bounded domain having a sectionally smooth boundary S, $S_T \equiv S \times (0,T)$, $Q_T \equiv \Omega \times (0,T)$, $\mathcal{D} \equiv \{\pi \equiv (u,v,w)\ :\ \pi \in \mathcal{D}_1 \times \mathcal{D}_2 \times \mathcal{D}_3\}$, $\mathcal{D}_1 \equiv \{u \in L_\infty^m(Q_T)\ :\ u(x,t) \in U$ for a.e. $(x,t) \in Q_T\}$, $\mathcal{D}_2 \equiv \{v \in L_\infty(\Omega) : v(x) \in V$ for a.e. $x \in \Omega\}$, $\mathcal{D}_3 \equiv \{w \in W_{2,1}^{0,1}(S_T)\ :\ \|w - \bar{w}\|_{2,1,S_T}^{(0,1)} \leq A\}$, and $\bar{w} \in W_{2,1}^{0,1}(S_T)$ is given.

Here and in what follows the following notation is used: $\|\varphi\|_{p,\Omega}$ is a norm in the space $L_p(\Omega)$ of functions $\varphi \colon \Omega \to R$ summable to p-th power (essentially bounded for $p = \infty$); $\|\cdot\|_{2,\Omega}^{(1)}$ is a norm in the space $W_2^1(\Omega)$; $|\cdot|_X^{(0)}$ is the standard norm in the space $C(X)$ of continuous functions $\varphi \colon X \to R$ on a compact set X; $M(X)$ is the set of all Radon measures on a compact set X, $\|\mu\|$ is the total variation of a measure $\mu \in M(X)$; $L_{2,1}(Q_T)$ is a Banach space of all Lebesgue measurable functions $\varphi \colon Q_T \to R$ such that the norm $\|\varphi\|_{2,1,Q_T} \equiv \int_0^T (\int_\Omega |\varphi(x,t)|^2 dx)^{1/2} dt$ is finite; $L_{2,1}(S_T)$ is a Banach space of all Lebesgue measurable functions $\varphi \colon S_T \to R$ such that the norm $\|\varphi\|_{2,1,S_T} \equiv \int_0^T (\int_S |\varphi(s,t)|^2 ds)^{1/2} dt$ is finite. By $W_{2,1}^{0,1}(S_T)$ denote the set of all functions $\varphi \in L_{2,1}(S_T)$ such that $\varphi_t \in L_{2,1}(S_T)$. A norm in the space $W_{2,1}^{0,1}(S_T)$ is defined by $\|\varphi\|_{2,1,S_T}^{(0,1)} \equiv \|\varphi\|_{2,1,S_T} + \|\varphi_t\|_{2,1,S_T}$. By $C^r([0,T],Y)$, where Y

is a infinite–dimensional Banach space, denote the space of r times strongly continuously differentiable functions $\varphi : [0, T] \to Y$ for $r > 0$, and the space of strongly continuous functions $\varphi : [0, T] \to Y$ for $r = 0$. A norm in the space $C^r([0, T], Y)$ is defined by $|z|_Y^{(r)} \equiv \sum_{i=0}^{r} \max_{t \in [0, T]} \|z_{t^{(i)}}(t)\|_Y$. Let us put $C([0, T], Y) \equiv C^0([0, T], Y)$.

Consider the following parametric optimization problem:

$$I_0(\pi) \to \inf, \quad \pi \in \mathcal{D}, \quad I_1(\pi) \in \mathcal{M} + q, \quad q \in C(X) \text{ is a parameter}, \quad (P_q)$$

where $\mathcal{M}$ is a set of all continuous nonpositive functions on the compact set $X \subseteq [0, T]$, the functional $I_0 \colon \mathcal{D} \to R$ and the operator $I_1 \colon \mathcal{D} \to C(X)$ are defined by

$$I_0(\pi) \equiv \int_\Omega G(x, z[\pi](x, T)) \, dx,$$

$$I_1(\pi)(\tau) \equiv \int_\Omega \Phi(x, \tau, z[\pi](x, \tau), v) \, dx, \quad \tau \in [0, T],$$

$z[\pi] \in W_2^1(Q_T)$ is an unique generalized solution (see [34]) to a third initial–boundary value problem

$$z_{tt} - \frac{\partial}{\partial x_i}(a_{ij} z_{x_j} + a_i z) + a(x, t, z, u) + b_i z_{x_i} = 0, \quad (x, t) \in Q_T, \quad (1.1)$$

$$z|_{t=0} = \varphi(x), \quad z_t|_{t=0} = v(x), x \in \Omega, \quad \frac{\partial z}{\partial N} + \sigma(s, t)z = w(s, t), (s, t) \in S_T,$$

corresponding to a triple $\pi \equiv (u, v, w) \in \mathcal{D}$. Here $\frac{\partial z}{\partial N} \equiv (a_{ij}(x, t)z_{x_j} + a_i(x, t)z) \cos \alpha_i(x, t)$, and $\alpha_i(x, t)$ is an angle between an outwards normal to S_T and Ox_i–axis.

Assume that

a) functions a_{ij}, a_{ijt}, a_i, a_{it}, b_i, b_{it}, $i, j = \overline{1, n}$, are Lebesgue measurable on Q_T;

b) functions σ and σ_t are Lebesgue measurable on S_T;

c) function $a \colon Q_T \times R \times R^m \to R$, together with $\nabla_z a$, is measurable in the Lebesgue sense with respect to (x, t, z, u) and continuous with respect to (z, u) for a.e. $(x, t) \in Q_T$;

d) function $G: \Omega \times R \to R$, together with $\nabla_z G$, is measurable in the Lebesgue sense with respect to (x, z) and continuous with respect to z for a.e. $x \in \Omega$;

e) function $\Phi: \Omega \times [0, T] \times R \times R \to R$, together with gradients $\nabla_z \Phi$, $\nabla_v \Phi$, is measurable in the Lebesgue sense with respect to $(x, t, z, v) \in \Omega \times [0, T] \times R \times V$ and continuous with respect to $(t, z, v) \in [0, T] \times R \times V$ for a.e. $x \in \Omega$;

f) the following conditions are fulfilled:

$$a_{ij} = a_{ji}, \quad \varphi \in W_2^1(\Omega), \quad \nu_1 |\xi|^2 \leq a_{ij}(x, t)\xi_i \xi_j \leq \nu_2 |\xi|^2$$
$$\forall (x, t) \in Q_T \quad (\nu_1, \ \nu_2 > 0);$$
$$|a_{ij}(x, t)| + |a_{ijt}(x, t)| + |a_i(x, t)| + |a_{it}(x, t)| +$$
$$+ |b_i(x, t)| + |b_{it}(x, t)| \leq \nu_3 \forall (x, t) \in Q_T, \ i, j = \overline{1, n};$$
$$|\sigma(s, t)| + |\sigma_t(s, t)| \leq \nu_4 \ \forall (s, t) \in S_T;$$
$$|a(x, t, 0, u)| \leq K(x, t) \ \forall u \in U \text{ and for a.e. } (x, t) \in Q_T,$$
$$\text{where } K \in L_2(Q_T);$$
$$|\nabla_z a(x, t, z, u)| \leq K_0 \ \forall (x, t, z, u) \in Q_T \times R \times U;$$

g) the following inequalities are fulfilled:

$$|\nabla_z a(x, t, z_1, u) - \nabla_z a(x, t, z_2, u)| \leq K_1 |z_1 - z_2|$$
$$\forall (x, t, z_i, u) \in Q_T \times R \times U, \ i = 1, 2;$$
$$|\nabla_z G(x, z_1) - \nabla_z G(x, z_2)| \leq K_1 |z_1 - z_2| \ \forall (x, z_i) \in \Omega \times R, \ i = 1, 2$$
$$|G(x, z)| + |\nabla_z G(x, z)| + |\nabla_v G(x, z)| \leq K_2 [1 + |z|] \ \forall (x, z) \in \Omega \times R$$
$$|\nabla_z \Phi(x, t, z_1, v_1) - \nabla_z \Phi(x, t, z_2, v_2)| +$$
$$+ |\nabla_v \Phi(x, t, z_1, v_1) - \nabla_v \Phi(x, t, z_2, v_2)| \leq K_1 [|z_1 - z_2| + |v_1 - v_2|]$$
$$\forall (x, t, z_i, v_i) \in \Omega \times [0, T] \times R \times V, \ i = 1, 2;$$
$$|\Phi(x, t, z, v)| + |\nabla_z \Phi(x, t, z, v)| + |\nabla_v \Phi(x, t, z, v)| \leq K_2 [1 + |z|]$$
$$\forall (x, t, z, v) \in \Omega \times [0, T] \times R \times V;$$
$$|\Phi(x, t_1, z, v) - \Phi(x, t_2, z, v)| + |\nabla_z \Phi(x, t_1, z, v) - \nabla_z \Phi(x, t_2, z, v)| +$$
$$|\nabla_v \Phi(x, t_1, z, v) - \nabla_v \Phi(x, t_2, z, v)| \leq K_3 (|t_1 - t_2|)$$
$$\forall (x, t_i, z, v) \in \Omega \times [0, T] \times R \times V, \ i = 1, 2,$$

where $K_3: [0, T] \to [0, +\infty)$ is such that $\lim_{\tau \to +0} K_3(\tau) = K_3(0) = 0$.

By definition, put $\mathcal{D}_q^\varepsilon \equiv \{\pi \in \mathcal{D} : I_1(\pi)(\tau) - q(\tau) \leq \varepsilon, \tau \in X\}, \varepsilon \geq 0$, $\beta(q) \equiv \beta_{+0}(q) \equiv \lim_{\varepsilon \to +0} \beta_\varepsilon(q)$, where $\beta_\varepsilon(q) \equiv \{\inf_{\pi \in \mathcal{D}_q^\varepsilon} I_0(\pi)$, if $\mathcal{D}_q^\varepsilon \neq \emptyset; +\infty$, if $\mathcal{D}_q^\varepsilon = \emptyset\}, \varepsilon \geq 0$. The function $\beta \colon C(X) \to R \cup \{+\infty\}$ is called the value function of the problem (P_q). It is obvious that $\beta(q) \leq \beta_0(q) \; \forall \, q \in C(X)$, where $\beta_0 \colon C(X) \to R$ is a classic value function. Suppose $\beta(q) < +\infty$. According to [21], a minimizing approximate solution (m.a.s.) in the problem (P_q) is a sequence of triples $\pi^i \in \mathcal{D}, i = 1, 2, \ldots$, such that

$$I_0(\pi^i) \leq \beta(q) + \delta^i, \quad \pi^i \in \mathcal{D}_q^{\varepsilon^i}, \quad i = 1, 2, \ldots, \tag{1.2}$$

where $\delta^i, \varepsilon^i, i = 1, 2, \ldots, \delta^i, \varepsilon^i \to 0, i \to \infty$, are sequences of nonnegative numbers.

2. Main Results

In this section, we formulate main results of the chapter. These results are connected with necessary conditions for m.a.s. elements, and with normality, regularity and sensitivity properties. In the sequel, we use the following standard notation: $H(x, t, z, u, \eta) \equiv \eta a(x, t, z, u)$.

The following theorem gives us the necessary conditions for a m.a.s., which is referred to as the maximum principle for the m.a.s..

Theorem 2.1 *Let* $\pi^i \equiv (u^i, v^i, w^i) \in \mathcal{D}, i = 1, 2, \ldots$, *be a m.a.s. to the problem* (P_q); *then there exist a sequence of numbers* $\gamma^i \geq 0, i = 1, 2, \ldots$, $\gamma^i \to 0, i \to \infty$, *a sequence of nonnegative numbers* $\lambda^i, i = 1, 2, \ldots$, *and a sequence of nonnegative Radon measures* $\mu^i \in M(X), i = 1, 2, \ldots$, *where* μ^i *is concentrated on the set*

$$X_i \equiv \{\tau \in X : |I_1(\pi^i)(\tau) - q(\tau)| \leq \gamma^i\}, \tag{2.1}$$

such that the following conditions are fulfilled:

a) a nontriviality condition of Lagrange multipliers:

$$\pi^i \in \mathcal{D}_q^{\gamma^i}, \quad \lambda^i + \|\mu^i\| = 1; \tag{2.2}$$

b) a maximum condition with respect to u:

$$\int_{Q_T} [\max_{u' \in U} H(x, t, z[\pi^i](x, t), u', \eta[\pi^i, \lambda^i, \mu^i](x, t)) - \tag{2.3}$$

$$- H(x, t, z[\pi^i](x, t), u^i(x, t), \eta[\pi^i, \lambda^i, \mu^i](x, t))] \, dx dt \leq \gamma^i,$$

c) a transversality condition with respect to v:

$$\max_{v\in\mathcal{D}_2}\left\{\int_\Omega \eta[\pi^i,\lambda^i,\mu^i](x,0)(v^i(x)-v(x))dx+ \right. \tag{2.4}$$

$$\left. +\int_X \mu^i(d\tau)\int_\Omega \nabla_v\Phi(x,\tau,z[\pi^i](x,\tau),v^i(x))(v^i(x)-v(x))dx\right\}\le\gamma^i,$$

d) a transversality condition with respect to w:

$$\max_{w\in\mathcal{D}_3}\int_{S_T}\eta[\pi^i,\lambda^i,\mu^i](s,t)(w^i(s,t)-w(s,t))dsdt\le\gamma^i, i=1,2,\dots. \tag{2.5}$$

Here $\eta[\pi,\lambda,\mu]$ is a solution to the adjoint initial–boundary value problem

$$\eta_{tt}-\frac{\partial}{\partial x_i}(a_{ij}\eta_{x_j}+b_i\eta)+a_i\eta_{x_i}+\nabla_z a(x,t,z[\pi](x,t),u)\eta= \tag{2.6}$$

$$=\nabla_z\Phi(x,t,z[\pi](x,t),v(x))\mu(dt),\quad (x,t)\in Q_T,$$

$$\eta|_{t=T}=0,\ \ \eta_t|_{t=T}=-\lambda\nabla_z G(x,z[\pi](x,T)),\ \ x\in\Omega,\ \ \left[\frac{\partial\eta}{\partial\mathcal{N}'}+\sigma\eta\right]\bigg|_{S_T}=0,$$

where $\frac{\partial\eta}{\partial\mathcal{N}'}\equiv(a_{ij}(x,t)\eta_{x_j}+b_i(x,t)\eta)\cos\alpha_i(x,t)$.

Remark 2.1 *From theorem 2.1 it follows that if a control $\pi^0\in\mathcal{D}_q^0$ is such that $I_0(\pi^0)=\beta(q)$, then this control satisfies the ordinary maximum principle for $\pi^i\equiv\pi^0$, $\gamma^i\equiv 0$, $(\lambda^i,\mu^i)\equiv(\lambda,\mu)$, $i=1,2,\dots$*

Further, like in [13], [14], we introduce the following natural definitions.

Definition 2.1 *A sequence $\pi^i\in\mathcal{D}$, $i=1,2,\dots$, is said to be a stationary sequence for the problem (P_q) if there exist a sequence of nonnegative numbers $\gamma^i\to 0$, $i\to\infty$, and a bounded sequence of pairs (λ^i,μ^i), $\lambda^i\ge 0$, where $\mu^i\in M[0,T]$ is nonnegative Radon measure such that μ^i is concentrated on the set X_i, $\pi^i\in\mathcal{D}_q^{\gamma^i}$, $i=1,2,\dots$, inequalities (2.3)–(2.5) are hold, and all limit points of the pair sequence (λ^i,μ^i), $i=1,2,\dots$, (in the *-weak sense for the second component) are not equal to zero.*

Definition 2.2 *A sequence $\pi^i\in\mathcal{D}$, $i=1,2,\dots$, stationary in the problem (P_q) is called normal, if all limit point of any corresponding sequence λ^i, $i=1,2,\dots$, are not equal to zero. The problem (P_q) is said to be normal if all stationary sequences of this problem are normal.*

A sequence $\pi^i\in\mathcal{D}$, $i=1,2,\dots$, stationary in the problem (P_q) is called regular, if all limit point of some corresponding sequence λ^i, $i=1,2,\dots$, are *not equal to zero. The problem (P_q) is said to be regular if there exists regular stationary sequence of this problem.*

Now we formulate regularity and normality conditions for the problem (P_q) (see, for example, [13], [14], [17], [18]).

Theorem 2.2 *Suppose the problem (P_q) is linearly convex, i.e., the function $a(x,t,z,u)$ has a form $a(x,t,z,u) \equiv a^1(x,t)z + a^2(x,t,u)$, the function G is convex with respect to z for a.e. $x \in \Omega$, and the function Φ is convex with respect to (z,v) for a.e. $x \in \Omega$ and for all $\tau \in [0,T]$. Then if the problem (P_q) is regular, then any stationary sequence is m.a.s..*

Theorem 2.3 *Let the problem (P_q) is linearly convex, and let one of the following conditions hold: 1) (Slater condition) there exists $\pi_0 \equiv (u_0, v_0, w_0) \in \mathcal{D}$ such that $I_1(\pi_0)(\tau) < q(\tau) \; \forall \tau \in X$; 2) (linearity condition) functions G and Φ have forms $G(x,z) \equiv G_1(x)z + G_2(x)$, $\Phi(x,\tau,z,v) \equiv \Phi_1(x,\tau)z + \Phi_2(x,\tau)v + \Phi_3(x,\tau)$, and there exists a nonstationary sequence $\pi^i \in \mathcal{D}_q^{\gamma^i}$, $\gamma^i \to 0$, $i \to \infty$, in the problem (P_q). Then the problem (P_q) is normal.*

By definition of the function $\beta_0 : C(X) \to R$, a quantity $\beta_0(q)$ is the classical lower bound in the problem (P_q). In the other words, $\beta_0(q)$ is the lower bound on the admissible controls class $\mathcal{D}$, if constraints are fulfilled in the classical exact sense. By this reason, the following question arises: is the length of the segment $[\beta(q), \beta_0(q)]$ equal to zero? The following two theorems connect the answer to this question with the problem (P_q) normality.

Theorem 2.4 *If $\beta(q) < \beta_0(q)$, then any sequence of triples $\pi^i \equiv (u^i, v^i, w^i)$, $i = 1, 2, \ldots$, such that*

$$I_0(\pi^i) \to \bar{\beta} \in [\beta(q), \beta_0(q)], \quad I_0(\pi^i) \leq \beta_0(q) + \varepsilon^i, \quad \pi^i \in \mathcal{D}_q^{\varepsilon^i},$$
$$\varepsilon^i \geq 0, \quad \varepsilon^i \to 0, \quad i \to \infty,$$

is stationary sequence. In addition, if $\bar{\beta} \in [\beta(q), \beta_0(q))$, then $\pi^i \equiv (u^i, v^i, w^i)$, $i = 1, 2, \ldots$, is not normal stationary sequence in the problem (P_q).

¿From this theorem it follows that

Theorem 2.5 *If the problem (P_q) is normal, then $\beta(q) = \beta_0(q)$.*

If the problem (P_q) is normal, then the following result about a stability of the problem (P_q) optimal value holds.

Theorem 2.6 *If the problem (P_q) is normal, then its value function β is Lipschitz in some neighborhood of the point $q \in C(X)$.*

In some sense, the following result is inverse to theorem 2.6.

Theorem 2.7 *Suppose the value function β of the problem (P_q) is Lipschitz in some neighborhood of the point q; then there exist regular minimizing approximate solutions in problem $(P_{q'})$ for all q' from this neighborhood.*

In the general case, when the value function $\beta(q)$ is not Lipschitz, the following general result is valid (which can be interpreted to mean that the set of regular (P_q) problems is "rather rich").

Theorem 2.8 *For any point $q \in \mathrm{dom}\ \beta$, any continuous positive function $\xi \in C(X)$, and for almost all points q' on the ray $\{q + t\xi : t \geq 0\}$, all minimizing approximate solution in problem $(P_{q'})$ are regular; i.e., the property according to which any m.a.s. in problem $(P_{q+t\xi})$ with given q and ξ is regular is general–position property for $t \geq 0$.*

3. Auxiliary Results

3.1. Main Equation

First of all, we need results concerning the main initial–boundary value problem (1.1). Let us formulate these results. Let us introduce some notation. For any triples $\pi^i \equiv (u^i, v^i, w^i) \in \mathcal{D}, i = 1, 2$, by definition, put $\mathcal{R}(u^1, u^2) \equiv \{(x, t) \in Q_T : u^1(x, t) \neq u^2(x, t)\}, d(\pi^1, \pi^2) \equiv \|v^1 - v^2\|_{\infty,\Omega} + meas\mathcal{R}(u^1, u^2) + \|w^1 - w^2\|_{2,1,S_T}^{(0,1)}$. Equip the set $\mathcal{D}$ with the metric $d(\cdot, \cdot)$. Then $\mathcal{D}$ is a complete metric space (see, e.g., [33]).

Let us introduce the following

Definition 3.1 *[34] A function $z = z[\pi] \in W_2^1(Q_T)$ is called a generalized solution to the initial–boundary value problem (1.1), corresponding to a triple $\pi \equiv (u, v, w) \in \mathcal{D}$, if z satisfies to the integral identity*

$$\int_{Q_T} [-z_t \eta_t + a_{ij} z_{x_j} \eta_{x_i} + a_i z \eta_{x_i} + b_i z_{x_i} \eta + a(x, t, z, u(x, t))\eta] dx dt +$$

$$+ \int_{S_T} \sigma z \eta ds dt = \int_{S_T} w \eta ds dt + \int_{\Omega} v(x) \eta(x, 0) dx$$

$$\forall \eta \in \hat{W}_2^1(Q_T); \quad z(x, 0) = \varphi(x), \quad x \in \Omega,$$

where $\hat{W}_2^1(Q_T) \equiv \{\eta \in W_2^1(Q_T) : \eta(x, T) = 0, \ x \in \Omega\}$.
The following result holds.

Lemma 3.1 *For any triple $\pi \equiv (u, v, w) \in \mathcal{D}$ there exists an unique solution $z[\pi] \in W_2^1(Q_T)$ to the initial–boundary value problem (1.1), and there exists a constant $c_0 > 0$ such that*

$$\|z[\pi]\|_{2,Q_T}^{(1)} \leq c_0 \;\; \forall \pi \in \mathcal{D}. \tag{3.1}$$

This constant is independent of a triple $\pi \in \mathcal{D}$. Moreover, for any two triples $\pi^i \equiv (u^i, v^i, w^i) \in \mathcal{D}$, $i = 1, 2$, the following inequality is fulfilled

$$\|z[\pi^1] - z[\pi^2]\|_{2,Q_T}^{(1)} \leq c_1(p)\Big(\|v^1 - v^2\|_{\infty,\Omega} + \tag{3.2}$$

$$+ (meas\mathcal{R}(u^1, u^2))^{\frac{p-1}{2p}} + \|w^1 - w^2\|_{2,1,S_T}^{(0,1)} + \Big[\int\limits_{\mathcal{R}(u^1,u^2)} K^2(x, t)dxdt \Big]^{1/2} \Big),$$

where $c_1(p) > 0$ is a constant. This is independent of triples $\pi^1, \pi^2 \in \mathcal{D}$; $p \in (1, +\infty)$ for $n = 1$, and $p \in (1, 1 + \frac{2}{n-1})$ for $n > 1$.

Proof. The first assertion of the lemma immediately follows from assumptions on input data of problem (P_q) and the a priori estimation of [34]. Namely, using methods of [35, Ch. IV, §3], it can be proved that there exists a constant $c_0^* > 0$ such that for all $\pi \equiv (u, v, w) \in \mathcal{D}$

$$\|z[\pi]\|_{2,Q_T}^{(1)} \leq c_0^*[\|\varphi\|_{2,\Omega}^{(1)} + \|v\|_{2,\Omega} + \|w\|_{2,1,S_T}^{(0,1)} + \|a(\cdot, \cdot, 0, u(\cdot, \cdot))\|_{2,1,Q_T}].$$

The constant c_0^* depends only on $\nu_1, \nu_2, \nu_3, \nu_4, T > 0$, and $K_0 > 0$. Using assumptions on input data of problem (P_q), we get estimation (3.1).

To prove the second assertion, let us linearize the problem (1.1):

$$\Delta z_{tt} - \frac{\partial}{\partial x_i}(a_{i,j}\Delta z_{x_j} + a_i\Delta z) + b_i\Delta z_{x_i} +$$

$$+ \Big[\int\limits_0^1 \nabla_z a(x, t, z_2 + \gamma\Delta z, u^2)d\gamma\Big]\Delta z = -\Delta_u a(x, t, z_1; u^1, u^2);$$

$$\Delta z(x, 0) = 0, \Delta z_t(x, 0) = \Delta v, x \in \Omega, \Big[\frac{\partial(\Delta z)}{\partial N} + \sigma\Delta z\Big]\Big|_{S_T} = \Delta w(s, t),$$

where $z_i \equiv z[\pi^i]$, $i = 1, 2$, $\Delta z \equiv z_1 - z_2$, $\Delta v \equiv v^1 - v^2$, $\Delta w \equiv w^1 - w^2$, $\Delta_u a(x, t, z; u^1, u^2) \equiv a(x, t, z, u^1) - a(x, t, z, u^2)$. Applying the above–mentioned a priori estimate, we obtain that

$$\|\Delta z\|_{2,Q_T}^{(1)} \leq c_0^*[\|\Delta v\|_{2,\Omega} + \|\Delta w\|_{2,1,S_T}^{(0,1)} +$$

$$+ \|\Delta_u a(\cdot, \cdot, z_1(\cdot, \cdot); u^1(\cdot, \cdot), u^2(\cdot, \cdot))\|_{2,1,Q_T}].$$

Using an embedding $W_2^1(\Omega)$ into $L_p(\Omega)$ for $p \in [1, \frac{2n}{n-2})$, and assumptions on input data of problem (P_q), we get (3.2).

According to [35, Ch. 1, Theorem 6.3 and Theorem 7.2] and embedding theorems for functions of one variable, the following lemma holds.

Lemma 3.2 *Suppose* $z \in W_2^1(Q_T)$; *then for any* $t \in [0, T]$ *there exists the trace* $z(\cdot, t) \in L_2(\Omega)$. *Moreover, there exist constants* $c_2 \equiv (1 + (1/T))^{1/2}$ *and* $c_3 > 0$ *such that*

$$\|z(\cdot, t)\|_{2,\Omega} \le c_2 \|z\|_{2,Q_T}^{(1)} \ \forall t \in [0, T]; \quad \|z\|_{2,S_T} \le c_3 \|z\|_{2,Q_T}^{(1)}.$$

Here $c_3 > 0$ *depends only on the domain* Ω *and on dimension* n. *Furthermore,*

$$\|z(\cdot, t_1) - z(\cdot, t_2)\|_{2,\Omega} \le \|z\|_{2,Q_T}^{(1)} |t_1 - t_2|^{1/2} \ \forall t_1, \ t_2 \in [0, T].$$

The following result follows from lemma 3.1 and lemma 3.2.

Lemma 3.3 *Suppose sequences of triples* $\pi^{i,k} \equiv (u^{i,k}, v^{i,k}, w^{i,k}) \in \mathcal{D}$, $k = 1, 2$, $i = 1, 2, \ldots$, *are such that* $\lim_{i \to \infty} d(\pi^{i,1}, \pi^{i,2}) = 0$; *then*

$$\lim_{i \to \infty} \|z[\pi^{i,1}] - z[\pi^{i,2}]\|_{2,Q_T}^{(1)} = 0, \ \lim_{i \to \infty} \|z[\pi^{i,1}] - z[\pi^{i,2}]\|_{2,S_T}^{(1)} = 0,$$

$$\lim_{i \to \infty} \max_{t \in [0,T]} \|z[\pi^{i,1}](\cdot, t) - z[\pi^{i,2}](\cdot, t)\|_{2,\Omega} = 0.$$

Finally, from lemmas 3.1–3.3 it follows that the following lemma holds.

Lemma 3.4 *The functional* $I_0 : \mathcal{D} \to R$ *and the operator* $I_1 : \mathcal{D} \to C(X)$ *are uniformly continuous and uniformly bounded on the complete metric space* $\mathcal{D}$. *Moreover, the set* $\{I_1(\pi) : \pi \in \mathcal{D}\}$ *is precompact in the space* $C(X)$.

3.2. Adjoint Equations

3.2.1. Equations with Radon Measure in the Right–Hand Side Part: An Unique Existence of Solutions

Consider the following initial–boundary value problem:

$$\eta_{tt} - \frac{\partial}{\partial x_i}(a_{ij}\eta_{x_j} + b_i\eta) + a_i\eta_{x_i} + a\eta = \tag{3.3}$$

$$= f(x,t) + g(x,t)\mu(dt), \quad (x,t) \in Q_T,$$

$$\eta(x,T) = \varphi(x), \ \eta_t(x,T) = \psi(x), \ x \in \Omega, \ \left[\frac{\partial \eta}{\partial N'} + \sigma\eta\right]\Big|_{S_T} = \omega(s,t),$$

where $f \in L_{2,1}(Q_T)$, $g \in C([0,T], L_2(\Omega))$, $\psi \in L_2(\Omega)$, $\varphi \in W_2^1(\Omega)$, $\omega \in W_{2,1}^{0,1}(S_T)$, $\mu \in M[0,T]$, and coefficients a_{ij}, a_i, b_i, a, σ, i, $j = \overline{1,n}$, are such that

$$a_{ij} = a_{ji}, \ \nu_1|\xi|^2 \le a_{ij}(x,t)\xi^i\xi^j \le \nu_2|\xi|^2 \qquad (3.4)$$
$$\forall (x,t) \in Q_T, \ \xi \in R^n \ (\nu_1, \ \nu_2 > 0);$$
$$|a_{ij}(x,t)| + |a_i(x,t)| + |b_i(x,t)| + |a(x,t)| + |a_{ijt}(x,t)| + |a_{it}(x,t)| +$$
$$+|b_{it}(x,t)| \le \nu_3 \ \forall (x,t) \in Q_T, \ i,j = \overline{1,n};$$
$$|\sigma(s,t)| + |\sigma_t(s,t)| \le \nu_4 \ \forall (s,t) \in S_T.$$

Let us give the following

Definition 3.2 *[36] A function $\eta \in W_2^1(Q_T)$ is said to be a solution to the initial–boundary value problem (3.3), if the following integral identity is fulfilled:*

$$\int_{Q_T} [-\eta_t z_t + a_{ij}\eta_{x_j} z_{x_i} + b_i \eta z_{x_i} + a_i \eta_{x_i} z + a\eta z] \, dxdt + \int_{S_T} \sigma \eta z \, dsdt +$$
$$+ \int_\Omega \psi(x) z(x,T) \, dx = \int_{Q_T} f(x,t) z(x,t) \, dxdt +$$
$$+ \int_{[0,T]} \left[\int_\Omega g(x,t) z(x,t) \, dx \right] \mu(dt)$$
$$\forall z \in W_2^1(Q_T), \ z(x,0) = 0, \ x \in \Omega; \ \eta(x,T) = \varphi(x), \ x \in \Omega.$$

The following result holds.

Lemma 3.5 *[36] Under the above–mentioned conditions, there exist an unique solution $\eta \in W_{2,0}^1(Q_T)$ to the initial–boundary value problem (3.3), and a constant $\bar{c}_0 > 0$ such that*

$$\|\eta\|_{2,Q_T}^{(1)} \le \bar{c}_0[\|\varphi\|_{2,\Omega}^{(1)} + \|\psi\|_{2,\Omega} + \|f\|_{2,1,Q_T} + \max_{t\in[0,T]} \|g(\cdot,t)\|_{2,\Omega}\|\mu\| +$$
$$+ \|\omega\|_{2,1,S_T}^{(0,1)}].$$

The constant $\bar{c}_0 > 0$ depends only on T, ν_1, ν_2, ν_3, $\nu_4 > 0$, on the domain Ω and on the dimension n.

Proof. Let us describe the scheme of the proof. Firstly, we approximate (in the *–weak sense) the Radon measure μ by a sequence of Radon measures μ^k,

$k = 1, 2, \ldots$ Each of approximating measures is absolutely continuous with respect to Lebesgue measure. Then, for each $k = 1, 2, \ldots$, we write out initial–boundary value problem (3.3), where μ is replaced by μ^k. Using results of [34], and, then, going to the limit as $k \to \infty$, we obtain the required results. For details, see [36].

3.2.2. Equations with Radon Measure in the Right–Hand Side Part: Integral Representation of the Solution

In this section, we get a special integral representation of the solution to a homogeneous third initial–boundary value problem for a linear divergent hyperbolic partial differential equation with Radon measure in the right–hand side part. To formulate this section result, let us introduce the following notation.

Suppose $t_1 < t_2$, $t_1, t_2 \in [0, T]$. By definition, put $Q_{(t_1,t_2)} \equiv \Omega \times (t_1, t_2)$, $Q_{[t_1,t_2]} \equiv \Omega \times [t_1, t_2]$, $S_{(t_1,t_2)} \equiv S \times (t_1, t_2)$. Suppose functions a_{ij}, a_i, b_i, $i, j = \overline{1, n}$, a, σ satisfy (3.4). Suppose $g \in C([0, T], L_2(\Omega))$, $\mu \in M[0, T]$. By definition, put $\chi(t, \tau) \equiv \{1, 0 \le t \le \tau \le T; \; 0, 0 \le \tau < t \le T\}$. Consider the following initial–boundary value problem:

$$f_{tt} - \frac{\partial}{\partial x_i}(a_{ij}f_{x_j} + b_i f) + a_i f_{x_i} + a f = g(x, t)\mu(dt), \quad (x, t) \in Q_T, \quad (3.5)$$

$$f(x, T) = f_t(x, T) = 0, \quad x \in \Omega, \quad \frac{\partial f}{\partial N'} + \sigma(s, t)f = 0, \quad (s, t) \in S_T.$$

By $f[a, g, \mu]$ denote the solution of this problem. Let us put $\mathfrak{p}[a, g](x, t, \tau) \equiv f[a, g, \delta_\tau](x, t)$, where δ_τ is a Radon δ–measure which concentrated at the point $t = \tau$.

Let us define a function $\mathfrak{x}[a, g](x, t, \tau)$, $(x, t) \in Q_T$, $\tau \in [0, T]$, for $(x, t) \in Q_{[0, \tau]}$ as a solution to the initial–boundary value problem

$$\mathfrak{x}_{tt} - \frac{\partial}{\partial x_i}(a_{ij}\mathfrak{x}_{x_j} + b_i\mathfrak{x}) + a_i\mathfrak{x}_{x_i} + a\mathfrak{x} = 0, \quad (x, t) \in Q_{[0, \tau]}, \quad (3.6)$$

$$\mathfrak{x}|_{t=\tau} = 0, \quad \mathfrak{x}_t|_{t=\tau} = -g(x, \tau), \quad x \in \Omega, \quad \frac{\partial \mathfrak{x}}{\partial N'} + \sigma(s, t)\mathfrak{x} = 0, \quad (s, t) \in S_{(0, \tau)},$$

and as a solution to the problem

$$\mathfrak{x}_{tt} - \frac{\partial}{\partial x_i}(a_{ij}\mathfrak{x}_{x_j} + b_i\mathfrak{x}) + a_i\mathfrak{x}_{x_i} + a\mathfrak{x} = 0, \quad (x, t) \in Q_{[\tau, T]}, \quad (3.7)$$

$$\mathfrak{x}|_{t=\tau} = 0, \quad \mathfrak{x}_t|_{t=\tau} = -g(x, \tau), \quad x \in \Omega, \quad \frac{\partial \mathfrak{x}}{\partial N'} + \sigma(s, t)\mathfrak{x} = 0, \quad (s, t) \in S_{(\tau, T)},$$

for $(x,t) \in Q_{[\tau, T]}$

The function $\mathfrak{x}[a,g]$ can be interpreted as a solution to the initial–boundary value problem

$$\mathfrak{x}_{tt} - \frac{\partial}{\partial x_i}(a_{ij}\mathfrak{x}_{x_j} + b_i\mathfrak{x}) + a_i\mathfrak{x}_{x_i} + a\mathfrak{x} = 0, \quad (x,t) \in Q_T, \qquad (3.8)$$

$$\mathfrak{x}|_{t=\tau} = 0, \quad \mathfrak{x}_t|_{t=\tau} = -g(x,\tau), \quad x \in \Omega, \quad \frac{\partial \mathfrak{x}}{\partial \mathcal{N}'} + \sigma(s,t)\mathfrak{x} = 0, \quad (s,t) \in S_T,$$

where initial conditions are given on a section of the cylinder Q_T by a hyperplane $t = \tau$.

In the connection with the selected research method of the considered optimal control problem, the question arises about an integral representation of the solution $\mathfrak{f}[a,g,\mu]$ to the initial–boundary value problem (3.5). Let us remember the scheme of this method. Firstly, we approximate the source problem (P_q) with pointwise state constraints by problems with finite number of functional constraints. Secondly, applying "standard" methods, we obtain "approximating" maximum principle in each "approximating" problem. Thirdly, we go to the limit in the family of approximating maximum principles as a number of functional inequality constraints tends to infinity. For this, we must "fuse" family of adjoint equations (each of adjoint equations corresponds to some functional inequality constraint) into one adjoint equation. The "fused" adjoint equation corresponds to the source state constraint, and involves a Radon measure in the right–hand side part. To provide this "fusion", we prove result on a representation of a solution to the problem (3.5) in the form of an integral of Green $\mathfrak{p}[a,g]$ function by measure μ with respect to τ. To prove this representation result, we need prove that the solution $\mathfrak{x}[a,g]$ of the problem (3.8) depends continuously on a position of the initial conditions hyperplane. Additionally, we need connection of Green function $\mathfrak{p}[a,g]$ of problem (3.5) with $\mathfrak{x}[a,g]$.

Theorem 3.1 *An inclusion $\mathfrak{x}[a,g] \in C([0,T], W_2^1(Q_T))$ holds, i.e., for any $\tau \in [0,T]$ there exists a trace $\mathfrak{x}[a,g](\cdot,\cdot,\tau) \in W_2^1(Q_T)$. This trace depends continuously on $\tau \in [0,T]$ in the norm of $W_2^1(Q_T)$. In addition,*

$$\mathfrak{p}[a,g](x,t,\tau) \equiv \mathfrak{x}[a,g](x,t,\tau)\chi(t,\tau), \quad (x,t) \in Q_T, \quad \tau \in [0,T]; \qquad (3.9)$$

$$\mathfrak{f}[a,g,\mu](x,t) = \int_{[0,T]} \mathfrak{p}[a,g](x,t,\tau)\mu(d\tau), \text{ for a.e. } (x,t) \in Q_T, \qquad (3.10)$$

and the following a priori estimate holds:

$$\max_{\tau \in [0,T]} \|\mathfrak{x}[a,g](\cdot,\cdot,\tau)\|_{2,Q_T}^{(1)} \leq \tilde{c} \max_{t \in [0,T]} \|g(\cdot,t)\|_{2,\Omega}. \qquad (3.11)$$

Here the constant $\tilde{c} > 0$ depends only on numbers T, ν_1, ν_2, $\nu_3 > 0$, on the domain Ω, and on a dimension n.

To prove this theorem, we need two lemmas. First of them is the consequence of [35, Ch.1, Theorem 6.1] and [37, p.157, Theorem 2.2].

Lemma 3.6 *Suppose $\Omega \subset R^n$ is a bounded domain with the sectionally smooth boundary; then there exists an orthonormal (in the space $L_2(\Omega)$) sequence $h_k \in W_2^1(\Omega)$, $k = 1, 2, \ldots$, such that for any $\varphi \in W_2^1(\Omega)$, $\psi \in L_2(\Omega)$, the following equalities are fulfilled:*

$$\lim_{N \to \infty} \|\varphi^N - \varphi\|_{2,\Omega}^{(1)} = 0, \quad \lim_{N \to \infty} \|\psi^N - \psi\|_{2,\Omega} = 0.$$

Here $\varphi^N(x) \equiv \sum_{k=1}^{N} \varphi_k h_k(x)$, $\psi^N(x) \equiv \sum_{k=1}^{N} \psi_k h_k(x)$, $\varphi_k \equiv \int_\Omega \varphi h_k \, dx$, $\psi_k \equiv \int_\Omega \psi h_k \, dx$, k, $N = 1, 2, \ldots$.

Lemma 3.7 *Suppose $h_k \in W_2^1(\Omega)$, $k = 1, 2, \ldots$, be an orthonormal (in the space $L_2(\Omega)$) sequence from the previous lemma; then for any functions $g_0 \in C^r([0, T], L_2(\Omega))$, $g_1 \in C^r([0, T], W_2^1(\Omega))$ ($r \geq 0$ is a fixed integer),*

$$\lim_{N \to \infty} |g_0^N - g_0|_{L_2(\Omega)}^{(0)} = 0, \quad \lim_{N \to \infty} |g_1^N - g_1|_{W_2^1(\Omega)}^{(r)} = 0,$$

where

$$g_0^N(x, t) \equiv \sum_{k=1}^{N} g_{0k}(t) h_k(x), \quad g_1^N(x, t) \equiv \sum_{k=1}^{N} g_{1k}(t) h_k(x),$$

$$g_{0k}(t) \equiv \int_\Omega g_0(x, t) h_k(x) dx, \quad g_{1k}(t) \equiv \int_\Omega g_1(x, t) h_k(x) dx,$$

$$k, N = 1, 2, \ldots, \quad t \in [0, T].$$

Moreover, the set $C^2([0, T], W_2^1(\Omega))$ is dense in the space $C([0, T], L_2(\Omega))$.

Proof. The proof of the lemma's first assertion is analogous to the proof of [37, Ch. IV, §4, Lemma 4.1]. By this reason, the proof of the first assertion is omitted. The second assertion follows immediately from the first assertion and the classic Weierstrass theorem about uniform approximation of continuous functions by algebraic polynoms. Lemma is proved.

The proof of Lemma 3.1. The proof is in three steps.

Step 1. Let us show that the equality (3.9) holds. First of all, according to results of [34], the function $\mathfrak{r}[a,g]$ is uniquely determined in the cylinder Q_T, for any $\tau \in [0, T]$. Besides, in view of Definition 3.2, the identities hold

$$\int_{Q_T} [-\mathfrak{p}_t[a,g](x,t,\tau)z_t + a_{ij}\mathfrak{p}_{x_j}[a,g](x,t,\tau)z_{x_i} + \qquad (3.12)$$

$$+ b_i\mathfrak{p}[a,g](x,t,\tau)z_{x_i} + + a_i\mathfrak{p}_{x_j}[g](x,t,\tau)z + a\mathfrak{p}[g](x,t,\tau)z]\,dxdt +$$

$$+ \int_{S_T} \sigma(s,t)\mathfrak{p}[a,g](x,t,\tau)z(s,t)dsdt = \int_{\Omega} g(x,\tau)z(x,\tau)\,dx$$

$$\forall\, z \in W_2^1(Q_T),\ z(\cdot, 0) \equiv 0;$$

$$\mathfrak{p}[a,g](x,T,\tau) \equiv 0,\ x \in \Omega,\ \tau \in [0, T].$$

In these identities, let us consider three cases.

Case A. Suppose $\tau = 0$; then the function $\mathfrak{p}[a,g](x,t,0)$, $(x,t) \in Q_T$, is the solution to the homogeneous initial–boundary value problem

$$\mathfrak{p}_{tt} - \frac{\partial}{\partial x_i}(a_{ij}\mathfrak{p}_{x_j} + b_i\mathfrak{p}) + a_i\mathfrak{p}_{x_i} + a\mathfrak{p} = 0,\ (x,t) \in Q_T,$$

$$\mathfrak{p}(x,T) = \mathfrak{p}_t(x,T) = 0,\ x \in \Omega,\ \frac{\partial\mathfrak{p}}{\partial N'} + \sigma(s,t)\mathfrak{p} = 0,\ (x,t) \in S_T.$$

According to [34], this problem has only trivial solution. Hence, in the case $\tau = 0$, the equality (3.9) holds.

Case B. Suppose $\tau = T$; then the function $\mathfrak{p}[a,g](\cdot,\cdot,T)$ coincides identically with the function $\mathfrak{r}[a,g](\cdot,\cdot,T)$ whence the equality (3.9) holds, if $\tau = T$.

Case C. Suppose $\tau \in (0, T)$. In the identities (3.12), let z be a linear combination of the form $\sum_{k=1}^{N} c_k(t)h_k(x)$, where N is some natural number, and functions c_k, $k = \overline{1, N}$, are piecewise differentiable on $[0, T]$, $c_k|_{[0,\tau]} \equiv 0$, $k = \overline{1, N}$. Here $h_k \in W_2^1(\Omega)$, $k = 1, 2, \ldots$, is the sequence from Lemma 3.6. Because the set of restrictions of all such z is everywhere dense in the set of all functions belonging to $W_2^1(Q_{(\tau,T)})$ and such that $z|_{t=\tau} = 0$, we conclude that

$$\int_{Q_{(\tau,T)}} [-\mathfrak{p}_t[a,g](x,t,\tau)z_t + a_{ij}\mathfrak{p}_{x_j}[a,g](x,t,\tau)z_{x_i} + b_i\mathfrak{p}[a,g](x,t,\tau)z_{x_i} +$$

$$+a_i\mathfrak{p}_{x_j}[a,g](x,t,\tau)z + a\mathfrak{p}[a,g](x,t,\tau)z]\,dxdt+$$

$$+\int_{S_{(\tau,T)}} \sigma\mathfrak{p}[a,g](s,t,\tau)z\,dsdt = 0,$$

$$\forall z \in W_2^1(Q_{(\tau,T)}), \quad z(\cdot,\tau) \equiv 0; \quad \mathfrak{p}[a,g](x,T,\tau) \equiv 0, \quad x \in \Omega.$$

Therefore, if $(x,t) \in Q_{(\tau,T)}$, then the function $\mathfrak{p}[a,g](\cdot,\cdot,\tau)$ is the solution to the homogeneous initial–boundary value problem

$$\mathfrak{p}_{tt} - \frac{\partial}{\partial x_i}(a_{ij}\mathfrak{p}_{x_j} + b_i\mathfrak{p}) + a_i\mathfrak{p}_{x_i} + a\mathfrak{p} = 0, (x,t) \in Q_{(\tau,T)},$$

$$\mathfrak{p}(x,T) = \mathfrak{p}_t(x,T) = 0, \quad x \in \Omega, \frac{\partial \mathfrak{p}}{\partial \mathcal{N}'} + \sigma(s,t)\mathfrak{p} = 0, (s,t) \in S_{(\tau,T)}.$$

According to [34], this problem has only trivial solution whence $\mathfrak{p}[a,g](x,t,\tau)$ $\equiv 0$, $(x,t) \in Q_{(\tau,T)}$. Hence, the relations (3.12) can be rewritten in the form

$$\int_{Q_{(0,\tau)}} [-\mathfrak{p}_t[a,g](x,t,\tau)z_t + a_{ij}\mathfrak{p}_{x_j}[a,g](x,t,\tau)z_{x_i} + b_i\mathfrak{p}[a,g](x,t,\tau)z_{x_i} +$$

$$+a_i\mathfrak{p}_{x_j}[g](x,t,\tau)z + a\mathfrak{p}[g](x,t,\tau)z]\,dxdt + \int_{S_{(0,\tau)}} \sigma\mathfrak{p}[a,g](s,t,\tau)z\,dsdt =$$

$$= \int_\Omega g(x,\tau)z(x,\tau)\,dx \quad \forall z \in W_2^1(Q_T), \quad z(\cdot,0) \equiv 0 \quad \tau \in [0,T];$$

$$\mathfrak{p}[a,g](x,\tau,\tau) \equiv 0, \quad x \in \Omega, \quad \tau \in [0,T].$$

In view of the last equalities, if $(x,t) \in Q_{(0,\tau)}$, then the function $\mathfrak{p}[a,g](\cdot,\cdot,\tau)$ is the solution to the initial–boundary value problem

$$\mathfrak{p}_{tt} - \frac{\partial}{\partial x_i}(a_{ij}\mathfrak{p}_{x_j} + b_i\mathfrak{p}) + a_i\mathfrak{p}_{x_i} + a\mathfrak{p} = 0, \quad (x,t) \in Q_{(0,\tau)},$$

$$\mathfrak{p}(x,\tau) = 0, \quad \mathfrak{p}_t(x,\tau) = -g(x,\tau), \quad x \in \Omega, \quad \left[\frac{\partial \mathfrak{p}}{\partial \mathcal{N}'} + \sigma(s,t)\mathfrak{p}\right]\bigg|_{S_{(0,\tau)}} = 0.$$

Thus, in the case $\tau \in (0,T)$, the equality (3.9) holds. Hence, (3.9) is completely proved.

Step 2. Using the Galerkin method, let us show an inclusion $\mathfrak{x}[a,g] \in C([0,T], W_2^1(Q_T))$. Suppose $g \in C^2([0,T], W_2^1(\Omega))$. Let $h_k \in W_2^1(\Omega)$, $k = 1, 2, \ldots$, be the sequence from Lemma 3.6, and let $g^N \equiv \sum\limits_{m=1}^{N} g_m(\tau)h_m(x)$, $g_m(\tau) \equiv \int_\Omega g(x,\tau)h_m(x)\,dx$, $m = \overline{1, N}$, $N = 1, 2, \ldots$, $\tau \in [0, T]$. Then, by virtue of Lemma 3.6,

$$\lim_{N\to\infty} |g^N - g|_{W_2^1(\Omega)}^{(2)} = 0. \tag{3.13}$$

We will find an approximation $\mathfrak{x}^N[a,g]$ to the function $\mathfrak{x}[a,g]$ in the form $\mathfrak{x}^N[a,g](x,t,\tau) \equiv \sum\limits_{k=1}^{N} e_k^N(t,\tau)h_k(x)$, where a collection of functions $e_k^N \in C^1(\Delta)$, $\Delta \equiv [0, T] \times [0, T]$, $k = \overline{1, N}$, is a unique continuous solution to the integral equation

$$e^N(t,\tau) + \int_t^\tau (y-t)\mathcal{A}^N(y)e^N(y,\tau)\,d\tau = \bar{g}^N(\tau)(\tau - t), \quad (t,\tau) \in \Delta. \tag{3.14}$$

Here $e^N(t,\tau) \equiv [e_1^N(t,\tau), \ldots, e_N^N(t,\tau)]^*$, $\bar{g}^N(\tau) \equiv [g_1(\tau), \ldots, g_N(\tau)]^* \in R^N$ are column–vectors. A matrix–valued function $\mathcal{A}^N : [0, T] \to R^{N\times N}$, $\mathcal{A}^N(t) = [\alpha_{km}(t)]_{k,m=\overline{1,N}}$, is such that

$$\alpha_{km}(t) \equiv \int_\Omega [a_{ij}(x,t)h_{kx_j}h_{mx_i} + b_i(x,t)h_k h_{mx_i} + a_i(x,t)h_{kx_i}h_m +$$

$$+a(x,t)h_k h_m]dx + \int_S \sigma(s,t)h_k(s)h_m(s)ds, \quad k, m = 1, 2, \ldots, \quad t \in [0, T].$$

Differentiating (3.14) with respect to t twice, we obtain that (3.14) is equivalent to the Cauchy problem

$$e_{tt}^N(t,\tau) + \mathcal{A}^N(t)e^N(t,\tau) = 0, \quad e^N(t,\tau)|_{t=\tau} = 0, \quad e_t^N(t,\tau)|_{t=\tau} = -\bar{g}^N(\tau),$$

Differentiating (3.14) with respect to τ once, and, then, with respect to t twice, we have

$$(e_\tau^N)_{tt} + \mathcal{A}^N(t)e_\tau^N = 0, \quad e_\tau^N|_{t=\tau} = \bar{g}^N(\tau), \quad (e_\tau^N)_t|_{t=\tau} = -\bar{g}^{N\prime}(\tau).$$

Applying reasoning similar to that in the proof of an energetic inequality from [35, Ch. IV, §3] and in the proof of an a priori estimate from [34], we conclude that

$$\|\mathfrak{x}^N[a,g](\cdot,\cdot,\tau_1) - \mathfrak{x}^N[a,g]](\cdot,\cdot,\tau_2)\|_{2,Q_T}^{(1)} \leq \tag{3.15}$$

$$\leq T\sqrt{C}|\tau_1 - \tau_2|^{1/2}|g^N|_{W_2^1(\Omega)}^{(1)},$$

$$\|\mathfrak{x}^N[a,g](\cdot,\cdot,\tau)\|_{2,Q_T}^{(1)} \leq \sqrt{CT}|g^N|_{L_2(\Omega)}^{(0)} \quad \forall \tau, \; \tau_1, \; \tau_2 \in [0, T].$$

Let $\tau_1, \tau_2 \in [0, T]$ be fixed. Applying reasoning similar to that in the proof of an existence of a solution to an initial–boundary value problem in [34], we obtain that there exist a subsequence N_m, $m = 1, 2, \ldots$, of the sequence $N = 1, 2, \ldots$, and functions $\mathfrak{x}[a,g](\cdot,\cdot,\tau_i)$, $i = 1, 2$, such that

$$\mathfrak{x}^{N_m}[a,g](\cdot,\cdot,\tau_i) \to \mathfrak{x}[a,g](\cdot,\cdot,\tau_i) \text{ weak in } W_2^1(Q_T), \tag{3.16}$$

$$\max_{t \in [0,T]} \|\mathfrak{x}^{N_m}[a,g](\cdot,t,\tau_i) - \mathfrak{x}[a,g](\cdot,t,\tau_i)\|_{2,\Omega} \to 0$$

$$\|\mathfrak{x}^{N_m}[a,g](\cdot,\cdot,\tau_i) - \mathfrak{x}[a,g](\cdot,\cdot,\tau_i)\|_{2,S_T} \to 0, \quad i = 1, 2, \; m \to \infty.$$

Moreover, each of functions $\mathfrak{x}[a,g](\cdot,\cdot,\tau_i)$, $i = 1, 2$, is a solution to initial–boundary value problem (3.6) for $(x,t) \in Q_{[0,\tau_i]}$, and is a solution to initial–boundary value problem (3.7) for $(x,t) \in Q_{[\tau_i,T]}$, where $\tau = \tau_i$, $i = 1, 2$. Using the limit relations (3.13) and (3.16), a weak compactness of a closed ball of Hilbert space, we get

$$\|\mathfrak{x}[a,g](\cdot,\cdot,\tau_1) - \mathfrak{x}[a,g](\cdot,\cdot,\tau_2)\|_{2,Q_T}^{(1)} \leq T\sqrt{C}|\tau_1 - \tau_2|^{1/2}|g|_{W_2^1(\Omega)}^{(2)}; \tag{3.17}$$

$$\|\mathfrak{x}[a,g](\cdot,\cdot,\tau)\|_{2,Q_T}^{(1)} \leq \sqrt{CT}|g|_{L_2(\Omega)}^{(0)}. \tag{3.18}$$

Thus, if $g \in C^2([0, T], W_2^1(\Omega))$, then $\mathfrak{x}[a,g] \in C([0, T], W_2^1(Q_T))$, and, additionally, $\mathfrak{x}[a,g]$ depends linearly on $g \in C^2([0, T], W_2^1(\Omega))$.

Using density $C^2([0, T], W_2^1(\Omega))$ in $C([0, T], L_2(\Omega))$, and above–mentioned linearly depending $\mathfrak{x}[a,g]$ on $g \in C^2([0, T], W_2^1(\Omega))$, we obtain that $\mathfrak{x}[a,g] \in C([0, T], W_2^1(Q_T))$ for all $g \in C([0, T], L_2(\Omega))$.

Step 3. Let us prove other assertions of the Lemma. From the inequality (3.18) and density $C^2([0, T], W_2^1(\Omega))$ in $C([0, T], L_2(\Omega))$ it follows that the estimation (3.11) holds. Since, by the second step of the proof, $\mathfrak{x}[a,g] \in$

$C([0, T], W_2^1(Q_T)) \: \forall \: g \in C([0, T], L_2(\Omega))$, then functions

$$\int_{[0,T]} \mathfrak{p}[a, g](x, t, \tau)\mu(d\tau), \quad \int_{[0,T]} \mathfrak{p}_t[a, g](x, t, \tau)\mu(d\tau),$$

$$\int_{[0,T]} \mathfrak{p}_{x_i}[a, g](x, t, \tau)\mu(d\tau), i = \overline{1, n}$$

are square summable over Q_T for any Radon measure $\mu \in M[0, T]$. Besides, using a definition of generalized derivatives in the sense of Sobolev, we conclude that

$$\frac{\partial}{\partial t} \int_{[0,T]} \mathfrak{p}[a, g](x, t, \tau)\mu(d\tau) = \int_{[0,T]} \mathfrak{p}_t[a, g](x, t, \tau)\mu(d\tau), \qquad (3.19)$$

$$\frac{\partial}{\partial x_i} \int_{[0,T]} \mathfrak{p}[a, g](x, t, \tau)\mu(d\tau) = \int_{[0,T]} \mathfrak{p}_{x_i}[a, g](x, t, \tau)\mu(d\tau),$$

$$i = \overline{1, n}, \; (x, t) \in Q_T.$$

Let us show that (3.10) take place. Indeed, because $\mathfrak{p}[a, g] \equiv \mathfrak{f}[a, g, \delta_\tau]$, then, writing out the corresponding integral identity, integrating this identity in $\tau \in [0, T]$ by a measure μ, and taking into account (3.19), we get that the function $\zeta(x, t) \equiv \int_0^T \mathfrak{p}[a, g](x, t, \tau)\mu(d\tau)$, $(x, t) \in Q_T$, is a solution to the problem (3.5). Because the problem (3.5) has an unique solution, then the equality (3.10) holds. Lemma 3.1 is completely proved.

Finally, let us prove the following result.

Lemma 3.8 *Let us consider the initial–boundary value problem*

$$z_{tt} - \frac{\partial}{\partial x_i}(a_{ij} z_{x_j} + a_i z) + b_i z_{x_i} + a z = f(x, t), \; (x, t) \in Q_T, \qquad (3.20)$$

$$z(x, 0) = 0, \; z_t(x, 0) = \psi(x), \; x \in \Omega, \; \frac{\partial z}{\partial N} + \sigma z = \omega(s, t), \; (s, t) \in S_T,$$

where coefficients a_{ij}, a_i, b_i, a, σ satisfy the conditions (3.4), and $f \in L_{2,1}(Q_T)$, $\psi \in L_2(\Omega)$, $\omega \in W_{2,1}^{0,1}(S_T)$.

Suppose function $z \in W_2^1(Q_T)$ is a solution to the problem (3.20), $\tau \in [0, T]$. Suppose a function $g : \Omega \times [0, T] \to R$ has the trace $g(\cdot, \tau) \in L_2(\Omega)$ for any $\tau \in [0, T]$, and this trace depends continuously on $\tau \in [0, T]$ in the norm *of the space $L_2(\Omega)$. Then*

$$\int_\Omega g(x,\tau)z(x,\tau)dx = \int_{Q_T} f(x,t)\mathfrak{p}[a,g](x,t,\tau)dxdt+ \qquad (3.21)$$

$$+ \int_\Omega \psi(x)\mathfrak{p}[a,g](x,0,\tau)dx + \int_{S_T} \mathfrak{p}[a,g](s,t,\tau)\omega(s,t)dsdt.$$

Proof. Because $z \in W_2^1(Q_T)$ is an unique solution to the problem (3.20), then the restrictions $z|_{Q_\tau}$ is an unique solution to the problem

$$z_{tt} - \frac{\partial}{\partial x_i}(a_{ij}z_{x_j} + a_i z) + b_i z_{x_i} + az = f(x,t), \quad (x,t) \in Q_\tau, \qquad (3.22)$$

$$z(x,0) = 0, \quad z_t(x,0) = \psi(x), \quad x \in \Omega, \quad \frac{\partial z}{\partial \mathcal{N}} + \sigma z = \omega(s,t), \quad (s,t) \in S_\tau.$$

In view of (3.22) we can write out the following identity which holds for all $\eta \in \hat{W}_2^1(Q_\tau)$:

$$\int_{Q_\tau}[-z_t\eta_t + a_{ij}z_{x_j}\eta_{x_i} + a_i z\eta_{x_i} + b_i z_{x_i}\eta + az\eta]dxdt + \int_{S_\tau} \sigma z\eta dsdt =$$

$$\qquad (3.23)$$

$$= \int_{S_\tau} \omega\eta dsdt + \int_{Q_\tau} f\eta dxdt + \int_\Omega \psi(x)\eta(x,0)dx; \quad z(x,0) = 0, \quad x \in \Omega.$$

However, in view of the lemma's conditions, all conditions of the unique existence theorem [34] of a solution $\mathfrak{x} \in W_2^1(Q_\tau)$ to the adjoint problem (3.6) are fulfilled whence the following identity holds for all $z \in W_2^1(Q_T)$, $z(\cdot,0) = 0$:

$$\int_{Q_\tau}[-z_t\mathfrak{x}_t + a_{ij}z_{x_j}\mathfrak{x}_{x_i} + a_i z\mathfrak{x}_{x_i} + b_i z_{x_i}\mathfrak{x} + az\mathfrak{x}]dxdt+ \qquad (3.24)$$

$$+ \int_{S_\tau} \sigma z\mathfrak{x}dsdt = \int_\Omega g(x,\tau)z(x,\tau)dx; \quad \mathfrak{x}(x,\tau) = 0, \quad x \in \Omega.$$

Substituting $\mathfrak{x}[a,g]$ for η in (3.23), and substituting the solution to the problem (3.22) for z in (3.24) and taking into account that left–hand side parts of (3.23) and (3.24) coincide, we conclude that

$$\int_\Omega g(x,\tau)z(x,\tau)dx = \int_{Q_\tau} f(x,t)\mathfrak{x}[a,g](x,t,\tau)dxdt+$$

$$+ \int_\Omega \psi(x)\mathfrak{x}[a,g](x,0,\tau)dx + \int_{S_\tau} \mathfrak{x}[a,g](s,t,\tau)\omega(s,t)dsdt.$$

The last equality can be rewritten in the form

$$\int_{\Omega} g(x, \tau) z(x, \tau) dx = \int_{Q_T} f(x, t) \mathfrak{r}[a, g](x, t, \tau) \chi(t, \tau) dx dt + \qquad (3.25)$$

$$+ \int_{\Omega} \psi(x) \mathfrak{r}[a, g](x, 0, \tau) \chi(0, \tau) dx + \int_{S_T} \mathfrak{r}[a, g](s, t, \tau) \chi(t, \tau) w(s, t) ds dt.$$

According to equality (3.1) from Lemma 3.9, we have that $\mathfrak{r}[a, g](x, t, \tau) \chi(t, \tau)$ $= \mathfrak{p}[a, g](x, t, \tau)$. Therefore, (3.25) can be rewritten in the form (3.21). The lemma is completely proved.

3.2.3. Adjoint Equations of Maximum Principle

In this section, we formulate facts concerning to a stability of solutions of maximum principle's adjoint equations under perturbations of controls. These facts will be used to prove main results. First of all, after investigation of special questions of the theory of linear hyperbolic equations with Radon measures in right–hand side parts, let us define Pontryagin maximum principle's adjoint functions of $\mathfrak{p}_0[\pi^1, \pi^2]$ and $\mathfrak{p}_1[\pi^1, \pi^2]$, $\pi^i \equiv (u^i, v^i, w^i) \in \mathcal{D}$, $i = 1, 2$, by

$$\mathfrak{p}_0[\pi^1, \pi^2](x, t) \equiv \mathfrak{p}[\mathfrak{a}[\pi^1, \pi^2], \mathfrak{g}_0[\pi^1, \pi^2]](x, t, T),$$

$$\mathfrak{p}_1[\pi^1, \pi^2](x, t, \tau) \equiv \mathfrak{p}[\mathfrak{a}[\pi^1, \pi^2], \mathfrak{g}_1[\pi^1, \pi^2]](x, t, \tau), \quad (x, t) \in Q_T, \quad \tau \in [0, T],$$

where

$$\mathfrak{a}[\pi^1, \pi^2](x, t) \equiv \int_0^1 \nabla_z a(x, t, z_2(x, t) + \gamma \Delta z(x, t), u^2(x, t)) d\gamma,$$

$$\mathfrak{g}_0[\pi^1, \pi^2](x) \equiv - \int_0^1 \nabla_z G(x, z_2(x, T) + \gamma \Delta z(x, T)) d\gamma,$$

$$\mathfrak{g}_1[\pi^1, \pi^2](x, \tau) \equiv - \int_0^1 \nabla_z \Phi(x, z_2(x, \tau) + \gamma \Delta z(x, \tau), v^2(x) + \gamma \Delta v(x)) d\gamma.$$

Here $z_i \equiv z[\pi^i]$, $i = 1, 2$, $\Delta z \equiv z_1 - z_2$, $\Delta v \equiv v^1 - v^2$. For brevity, set $\mathfrak{g}_0[\pi, \pi] \equiv \mathfrak{g}_0[\pi]$, $\mathfrak{g}_1[\pi, \pi] \equiv \mathfrak{g}_1[\pi]$, $\mathfrak{p}_0[\pi, \pi] \equiv \mathfrak{p}_0[\pi]$, $\mathfrak{p}_1[\pi, \pi] \equiv \mathfrak{p}_1[\pi]$, $\mathfrak{a}[\pi, \pi] \equiv \mathfrak{a}[\pi]$. Let us remember that $\mathfrak{p}[a, g](x, t, \tau) \equiv \mathfrak{f}[a, g, \delta_\tau](x, t)$, $(x, t) \in Q_T$, $\tau \in [0, T]$, where $\mathfrak{f}[a, g, \delta_\tau]$ is a solution to the problem (3.5) for $\mu = \delta_\tau$.

From assumptions on input data of problem (P_q), and from Theorem 3.1, it follows that the following lemma holds.

Lemma 3.9 *If sequences $\pi^{i,k} \in \mathcal{D}$, $k = \overline{1,4}$, $i = 1, 2, \ldots$, are such that* $\lim\limits_{i \to \infty} d(\pi^{i,1}, \pi^{i,3}) = 0$, $\lim\limits_{i \to \infty} d(\pi^{i,2}, \pi^{i,4}) = 0$, *then*

$$\lim_{i \to \infty} \|\mathfrak{p}_0[\pi^{i,1}, \pi^{i,2}] - \mathfrak{p}_0[\pi^{i,3}, \pi^{i,4}]\|_{2,Q_T}^{(1)} = 0,$$

$$\lim_{i \to \infty} \sup_{\tau \in [0,T]} \|\mathfrak{p}_1[\pi^{i,1}, \pi^{i,2}](\cdot, \cdot, \tau) - \mathfrak{p}_1[\pi^{i,3}, \pi^{i,4}](\cdot, \cdot, \tau)\|_{2,Q_T}^{(1)} = 0.$$

3.3.   A Calculation of First Variations

In the sequel, in a definition and a calculation of first variations of functionals, a notion of an iterative limit will play the main role. Besides, corresponding generalizations of a classic notion of Lebesgue point will also play the important role.

Definition 3.3[22] – [24] *Let $G \subset R^n$ be an open set. A point $x \in G$ is said to be an (l, m)–Lebesgue point of a summable function $f : G \to R^1$, $1 \leq l \leq m \leq n$, if $f(x) \neq \infty$ and*

$$\lim_{h \to 0} \frac{1}{(2h)^{m-l+1}} \int_{x_l-h}^{x_l+h} \cdots \int_{x_m-h}^{x_m+h} |f(x_1, \ldots, x_{l-1}, y_1, \ldots, y_{m-l+1}, x_{m+1}, \ldots, x_n) - $$
$$ - f(x)| \, dy_1 \ldots dy_{m-l+1} = 0.$$

Remark 3.1 *It is easy to see that an $(1, n)$–Lebesgue point is a Lebesgue point in an usual sense* [27].

Lemma 3.10[22] – [25] *For any fixed l, m, $1 \leq l \leq m \leq n$, almost all points of an open set G are (l, m)–Lebesgue points of a summable function $f : G \to R^1$.*

On the base of these notions and results, we calculate the first variations of functionals $I_0(\cdot)$ and $I_1(\cdot)(\tau)$ $\forall \tau \in [0, T]$. Let $\pi \equiv (u, v, w) \in \mathcal{D}$ be arbitrary. Let us construct a collection of variation parameters $\mathfrak{m} \equiv (\{(x^i, t^i), \gamma^{i,r}, u^{i,r}, i = \overline{1, i_1}, r = \overline{1, r_0(i)}\}, \tilde{v}, \tilde{w})$, where $(x^i, t^i) \in Q_T$, $\gamma^{i,r} \geq 0$, $i = \overline{1, i_1}$, $r = \overline{1, r_0(i)}$, $\sum\limits_{i=1}^{i_1} \sum\limits_{r=1}^{r_0(i)} \gamma^{i,r} \leq 1$, $\tilde{v} \in \mathcal{D}_2$, $\tilde{w} \in \mathcal{D}_3$, $u^{i,r} \in U^*$, $i = \overline{1, i_1}$, $r = \overline{1, r_0(i)}$; $U^* \subseteq U$ is a countable set, U^* is everywhere dense in U. Let us denote the set of all such collections $\mathfrak{m}$ by $\mathfrak{M}$.

A triple $\pi_\varepsilon \equiv (u_\varepsilon, v_\varepsilon, w_\varepsilon) \in \mathcal{D}$ is called a variation of triple $\pi \equiv (u, v, w) \in$

$\mathcal{D}$, where $\varepsilon \equiv (\varepsilon_1, \varepsilon_2)$, $\varepsilon_1, \varepsilon_2 \geq 0$, $0 \leq \varepsilon_1, \varepsilon_2 \leq \varepsilon_0 < 1$, if

$$u_\varepsilon(x, t) \equiv \begin{cases} u^{i,r}, & (x, t) \in Q_{i,r}^\varepsilon, \ i = \overline{1, i_1}, r = \overline{1, r_0(i)}; \\ u(x, t), & (x, t) \in Q_T \setminus \bigcup_{i=1}^{i_1} \bigcup_{r=1}^{r_0(i)} Q_{i,r}^\varepsilon; \end{cases}$$

$$v_\varepsilon(x) \equiv v(x) + \varepsilon_1^n \varepsilon_2(\tilde{v}(x) - v(x)) \equiv v(x) + \varepsilon_1^n \varepsilon_2 \delta v(x), \quad x \in \Omega;$$

$$w_\varepsilon(s, t) \equiv w(s, t) + \varepsilon_1^n \varepsilon_2(\tilde{w}(s, t) - w(s, t)) \equiv w(s, t) +$$
$$+ \varepsilon_1^n \varepsilon_2 \delta w(s, t), \quad (s, t) \in S_T;$$

where $Q_{i,r}^\varepsilon \equiv Q_{i,r}^{\varepsilon_1,\varepsilon_2} \equiv Q_{i,r,1}^{\varepsilon_1} \times Q_{i,r,2}^{\varepsilon_2}$, $Q_{i,r,1}^{\varepsilon_1} \equiv \prod_{\alpha=1}^n (x_\alpha^i - \varepsilon_1, x_\alpha^i - \varepsilon_1(r-1)]$,

$Q_{i,r,2}^{\varepsilon_2} \equiv (t^i - \varepsilon_2 \sum_{\alpha=1}^r \gamma^{i,\alpha}, t^i - \varepsilon_2 \sum_{\alpha=1}^{r-1} \gamma^{i,\alpha}]$. Here $\varepsilon_0 > 0$ is a small enough number depending on $\gamma^{i,r}$ and (x^i, t^i), $i = \overline{1, i_1}$, $r = \overline{1, r_0(i)}$, such that sets

$$Q_i^{\varepsilon_0} \equiv Q_{i,1}^{\varepsilon_0} \times Q_{i,2}^{\varepsilon_0} \equiv \prod_{\alpha=1}^n [x_\alpha^i - \varepsilon_0 r_0(i), x_\alpha^i] \times [t^i - \varepsilon_0 \sum_{\alpha=1}^{r_0(i)} \gamma^{i,\alpha}, t^i]), i = \overline{1, i_0(i)},$$

do not pairwise intersect.

Let us formulate the following obvious result.

Lemma 3.11 *There exists a constant $L \equiv 1 + measV + 2A$ such that for any $\pi \equiv (u, v, w) \in \mathcal{D}$*

$$d(\pi, \pi_\varepsilon) \leq L\varepsilon_1^n \varepsilon_2.$$

In the sequel, we need the following lemma (see, e.g., [25, Lemma 6])

Lemma 3.12 *Suppose $\mathcal{A}_1 \subset R^n$ and a segment $\mathcal{A}_2 \equiv [a_1, a_2] \subset R$ are such that the Lebesgue measure of $\mathcal{A}_1$ is positive and finite, and $\mathcal{A}_1 \times \mathcal{A}_2 \subset Q_T$. Then there exists a constant $C > 0$ such that*

$$\int_{\mathcal{A}_1} \|\zeta(x, \cdot)\|_{\infty, \mathcal{A}_2}^2 \, dx \leq [C\|\zeta\|_{2,Q_T}^{(1)}]^2 \quad \forall \zeta \in W_2^1(Q_T).$$

The constant $C > 0$ depends only on the length of the segment $\mathcal{A}_2$.

Taking into account this lemma, we will obtain the expressions for functionals I_0 and I_1 first variations.

Lemma 3.13 *1) For any $\pi \equiv (u, v, w) \in \mathcal{D}$ there exists a subset $Q_0[\pi] \subseteq Q_T$ such that $\text{meas} \, Q_0[\pi] = \text{meas} \, Q_T$ and for all $\mathfrak{m} \equiv (\{(x^i, t^i), \gamma^{i,r}, u^{i,r}, i = \overline{1, i_1}, r = \overline{1, r_0(i)}\}, \tilde{v}, \tilde{w}) \in \mathfrak{M}$, $(x^i, t^i) \in Q_0[\pi]$, $i = \overline{1, i_1}$, there exists a variation*

$$\delta I_0(\pi; \mathfrak{m}) \equiv \lim_{\varepsilon_1 \to +0} \frac{1}{\varepsilon_1^n} \lim_{\varepsilon_2 \to +0} \frac{I_0(\pi_{\varepsilon_1, \varepsilon_2}) - I_0(\pi)}{\varepsilon_2}.$$

In addition, the following representation holds

$$\delta I_0(\pi; \mathfrak{m}) \equiv \int_\Omega \mathfrak{p}_0[\pi](x, 0)\delta v(x)\, dx + \int_{S_T} \mathfrak{p}_0[\pi](s, t)\delta w(s, t)\, dsdt -$$

$$- \sum_{i=1}^{i_1} \sum_{r=1}^{r_0(i)} \gamma^{i,r} \mathfrak{p}_0[\pi](x^i, t^i)\Delta_u a(x^i, t^i, z[\pi](x^i, t^i); u^{i,r}, u(x^i, t^i)).$$

2) For any point $\tau \in [0, T]$ and any triple $\pi \equiv (u, v, w) \in \mathcal{D}$ there exists a subset $Q_1[\pi, \tau] \subseteq Q_T$ such that $\operatorname{meas} Q_1[\pi, \tau] = \operatorname{meas} Q_T$ and for all $\mathfrak{m} \equiv (\{(x^i, t^i), \gamma^{i,r}, u^{i,r}, i = \overline{1, i_1}, r = \overline{1, r_0(i)}\}, \tilde{v}, \tilde{w}) \in \mathfrak{M}$, $(x^i, t^i) \in Q_1[\pi, \tau]$, $i = \overline{1, i_1}$, there exists a variation

$$\delta I_1(\pi, \tau; \mathfrak{m}) \equiv \lim_{\varepsilon_1 \to +0} \frac{1}{\varepsilon_1^n} \lim_{\varepsilon_2 \to +0} \frac{I_1(\pi_{\varepsilon_1, \varepsilon_2})(\tau) - I_1(\pi)(\tau)}{\varepsilon_2}.$$

This variation can be represented in the form

$$\delta I_1(\pi, \tau; \mathfrak{m}) \equiv + \int_\Omega [\mathfrak{p}_1[\pi](x, 0, \tau) + \nabla_v \Phi(x, \tau, z[\pi](x, \tau), v)]\delta v(x)\, dx$$

$$+ \int_{S_T} \mathfrak{p}_1[\pi](s, t, \tau)\delta w\, dsdt -$$

$$- \sum_{i=1}^{i_1} \sum_{r=1}^{r_0(i)} \gamma^{i,r} \mathfrak{p}_1[\pi](x^i, t^i, \tau)\Delta_u a(x^i, t^i, z[\pi](x^i, t^i); u^{i,r}, u(x^i, t^i)).$$

Proof. Let us prove only second assertion of the lemma, because a calculation of the iterative limit $\delta I_0(\pi; \mathfrak{m})$ is completely analogous to a calculation of an analogous limit in the work [25]. Let a triple $\pi \equiv (u, v, w) \in \mathcal{D}$ be fixed. Linearizing the equation (1.1), and setting $z_\varepsilon \equiv z[\pi_\varepsilon]$, $z \equiv z[\pi]$, $\Delta_\varepsilon z \equiv z_\varepsilon - z$, $\Delta_\varepsilon v \equiv v_\varepsilon - v$, $\Delta_\varepsilon w \equiv w_\varepsilon - w$, we obtain that

$$\Delta_\varepsilon z_{tt} - \frac{\partial}{\partial x_i}(a_{ij}\Delta_\varepsilon z_{x_j} + a_i\Delta_\varepsilon z) + b_i\Delta_\varepsilon z_{x_j} + \mathfrak{a}[\pi_\varepsilon, \pi]\Delta_\varepsilon z =$$

$$= -\Delta_u a(x, t, z_\varepsilon; u_\varepsilon, u);$$

$$\Delta_\varepsilon z|_{t=0} = 0, \Delta_\varepsilon z_t|_{t=0} = \Delta_\varepsilon v, x \in \Omega, \frac{\partial(\Delta_\varepsilon z)}{\partial \mathcal{N}} + \sigma\Delta_\varepsilon z = \Delta_\varepsilon w, (s, t) \in S_T.$$

Linearizing the increment $\Delta_\varepsilon I_1(\tau) \equiv I_1(\pi_\varepsilon)(\tau) - I_1(\pi)(\tau)$ of the functional $I_1(\cdot)(\tau)$, we have

$$\Delta_\varepsilon I_1(\tau) = -\int_\Omega \mathfrak{g}_1[\pi_\varepsilon, \pi](x, \tau)\Delta_\varepsilon z(x, \tau)dx+$$

$$+\int_\Omega \left[\int_0^1 \nabla_v \Phi(x, \tau, z(x, \tau) + y\Delta_\varepsilon z(x, \tau), v(x) + y\Delta_\varepsilon v(x))dy\right]\Delta_\varepsilon v(x)dx.$$

Applying lemma 3.8 to the last equality, we obtain

$$\Delta_\varepsilon I_1(\tau) = \left\{\int_{Q_T} \mathfrak{p}_1[\pi_\varepsilon, \pi](x, t, \tau)[\Delta_u a(x, t, z_\varepsilon; u_\varepsilon, u) - \Delta_u a(x, t, z; u_\varepsilon, u)]dxdt\right\}+$$

$$+\left\{-\int_{Q_T} \Delta_\varepsilon \mathfrak{p}_1(x, t, \tau)\Delta_u a(x, t, z; u_\varepsilon, u)dxdt\right\}+$$

$$+\left\{-\int_{Q_T} \mathfrak{p}_1[\pi](x, t, \tau)\Delta_u a(x, t, z; u_\varepsilon, u)dxdt\right\}+\left\{\int_{S_T} \mathfrak{p}_1[\pi_\varepsilon, \pi](s, t, \tau)\Delta_\varepsilon w \, dsdt+\right.$$

$$\int_\Omega \left[\mathfrak{p}_1[\pi_\varepsilon, \pi](x, 0, \tau) + \int_0^1 \nabla_v \Phi(x, \tau, z(x, \tau) + y\Delta_\varepsilon z(x, \tau), v + y\Delta_\varepsilon v)dy\right]\Delta_\varepsilon v dx\right\} \equiv$$

$$\equiv \{\Delta_\varepsilon I_1^{(1)}(\tau)\} + \{\Delta_\varepsilon I_1^{(2)}(\tau)\} + \{\Delta_\varepsilon I_1^{(3)}(\tau)\} + \{\Delta_\varepsilon I_1^{(4)}(\tau)\}.$$

Here $\Delta_\varepsilon \mathfrak{p}_1 \equiv \mathfrak{p}_1[\pi_\varepsilon, \pi] - \mathfrak{p}_1[\pi]$.

Using assumptions on an integrand Φ, lemma 3.2, lemma 3.3, lemma 3.9, and the definition of a variation π_ε of controls, we conclude

$$\lim_{\varepsilon_1 \to +0} \frac{1}{\varepsilon_1^n} \lim_{\varepsilon_2 \to +0} \frac{\Delta_\varepsilon I_1^{(4)}(\tau)}{\varepsilon_2} = \int_{S_T} \mathfrak{p}_1[\pi](s, t, \tau)\delta w dsdt+ \qquad (3.26)$$

$$+\int_\Omega [\nabla_v \Phi(x, \tau, z(x, \tau), v) + \mathfrak{p}_1[\pi](x, 0, \tau)]\delta v dx.$$

Let us calculate similar limit for the summand $\Delta_\varepsilon I_1^{(1)}(\tau)$:

$$\left|\frac{1}{\varepsilon_2}\Delta_\varepsilon I_1^{(1)}(\tau)\right| \le \frac{1}{\varepsilon_2}\sum_{i=1}^{i_1}\sum_{r=1}^{r_0(i)} \int_{Q_{i,r,1}^{\varepsilon_1}} dx \int_{Q_{i,r,2}^{\varepsilon_2}} dt \int_0^1 |\nabla_z a(x, t, z + y\Delta_\varepsilon z, u_\varepsilon)-$$

$$-\nabla_z a(x,t,z+y\Delta_\varepsilon z, u)\|\mathfrak{p}_1[\pi_\varepsilon,\pi](x,t,\tau)\|\Delta_\varepsilon z|dy \leq$$

$$\leq \frac{2K_0}{\varepsilon_2} \sum_{i=1}^{i_1}\sum_{r=1}^{r_0(i)} \int_{Q_{i,r,1}^{\varepsilon_1}} \gamma^{i,r}\varepsilon_2\|\mathfrak{p}_1[\pi_\varepsilon,\pi](x,\cdot,\tau)\|_{\infty,Q_{i,2}^{\varepsilon_0}}\|\Delta_\varepsilon z(x,\cdot)\|_{\infty,Q_{i,2}^{\varepsilon_0}}\,dx \leq$$

$$\leq 2K_0 \sum_{i=1}^{i_1}\sum_{r=1}^{r_0(i)} \left[\int_{Q_{i,1}^{\varepsilon_0}} \|\mathfrak{p}_1[\pi_\varepsilon,\pi](x,\cdot,\tau)\|^2_{\infty,Q_{i,2}^{\varepsilon_0}}\,dx\right]^{1/2} \left[\int_{Q_{i,1}^{\varepsilon_0}} \|\Delta_\varepsilon z(x,\cdot)\|^2_{\infty,Q_{i,2}^{\varepsilon_0}}\,dx\right]^{1/2}.$$

Applying lemma 3.12 with $\mathcal{A}_1 \equiv Q_{i,1}^{\varepsilon_0}$ and $\mathcal{A}_2 \equiv Q_{i,2}^{\varepsilon_0}$ to expressions in square brackets, we get that there exists a constant $C > 0$ such that

$$\left|\frac{1}{\varepsilon_2}\Delta_\varepsilon I_1^{(1)}(\tau)\right| \leq 2K_0 C^2 \sum_{i=1}^{i_1}\sum_{r=1}^{r_0(i)} \|\mathfrak{p}_1[\pi_\varepsilon,\pi](\cdot,\cdot,\tau)\|_{2,Q_T}^{(1)}\|\Delta_\varepsilon z\|_{2,Q_T}^{(1)} \leq$$

$$\leq C_1\|\Delta_\varepsilon z\|_{2,Q_T}^{(1)}.$$

Hence, in view of Lemma 3.3 and the definition of a variation π_ε,

$$\lim_{\varepsilon_1\to+0}\frac{1}{\varepsilon_1^n}\lim_{\varepsilon_2\to+0}\frac{\Delta_\varepsilon I_1^{(1)}(\tau)}{\varepsilon_2} = 0. \tag{3.27}$$

Consider summands $\Delta_\varepsilon I_1^{(2)}(\tau)$ and $\Delta_\varepsilon I_1^{(3)}(\tau)$. Let

$$\mathfrak{F}_{i,r}(x,t) \equiv \begin{cases} -\Delta_u a(x,t,z(x,t);u_\varepsilon(x,t),u(x,t)), \ (x,t) \in Q_{i,r}^{\varepsilon_0}; \\ 0, \ (x,t) \in R^{n+1}\setminus Q_{i,r}^{\varepsilon_0}. \end{cases}$$

It is clear, that $\mathfrak{F}_{i,r} \in L_2(R^{n+1})$, $i = \overline{1, i_1}$, $r = \overline{1, r_0(i)}$. Let us introduce a maximal function

$$M_t\mathfrak{F}_{i,r}(x,t) \equiv \sup_{\delta>0}\frac{1}{2\delta}\int_{t-\delta}^{t+\delta} |\mathfrak{F}_{i,r}(x,y)|dy.$$

On the strength of the classic maximal function theorem (see, e.g., proposition c) of [27, p.15, Theorem 1], where a constant A_p depends only on p and n), $M_t\mathfrak{F}_{i,r} \in L_2(R^{n+1})$ whence

$$M_t\mathfrak{F}_{i,r}(\cdot,t) \in L_2(R^n) \text{ for a.e. } t \in R^1, \ i = \overline{1, i_1}, \ r = \overline{1, r_0(i)}. \tag{3.28}$$

Let us require that the following assumptions on points (x^i, t^i) are fulfilled:

1) points (x^i, t^i), $i = \overline{1, i_1}$, are $(1, n+1)$, $(1, n)$–Lebesgue points ($t \equiv x_{n+1}$) (see Definition 3.3) of all functions

$$-\mathfrak{p}_1[\pi](\cdot, \cdot, \tau)\Delta_u a(\cdot, \cdot, z(\cdot, \cdot); u', u(\cdot, \cdot)), \quad u' \in U^*; \qquad (3.29)$$

2) almost all points of sections $Q_T^{t^i} \equiv \{(x, t) \in Q_T : (x, t) = (x, t^i)\}$, $i = \overline{1, i_1}$, are $(n+1, n+1)$–Lebesgue points of all functions (3.29);

3) the following inclusions holds:

$$M_t \mathfrak{F}_{i,r}(\cdot, t^i) \in L_2(R^n), \quad i = \overline{1, i_1}, \quad r = \overline{1, r_0(i)}.$$

By virtue of Lemma 3.10 and inclusions (3.28), such selection of points (x^i, t^i) is possible, and, additionally, the Lebesgue measure of all such points set $Q_1[\pi, \tau]$ coincides with the Lebesgue measure of the cylinder Q_T.

Let us estimate a summand $\Delta_\varepsilon I_1^{(2)}(\tau)$:

$$\left| \frac{1}{\varepsilon_2} \Delta_\varepsilon I_1^{(2)}(\tau) \right| \leq$$

$$\leq \sum_{i=1}^{i_1} \sum_{r=1}^{r_0(i)} \int_{Q_{i,r,1}^{\varepsilon_1}} \left[\frac{1}{\varepsilon_2} \int_{t^i - \delta^{i,r}\varepsilon_2}^{t^i + \delta^{i,r}\varepsilon_2} |\mathfrak{F}_{i,r}(x, t)| dt \right] \|\Delta_\varepsilon \mathfrak{p}_1(x, \cdot, \tau)\|_{\infty, Q_{i,2}^{\varepsilon_0}} dx \leq$$

$$\leq \sum_{i=1}^{i_1} \sum_{r=1}^{r_0(i)} \int_{Q_{i,r,1}^{\varepsilon_2}} M_t \mathfrak{F}_{i,r}(x, t^i) 2\delta^{i,r} \|\Delta_\varepsilon \mathfrak{p}_1(x, \cdot, \tau)\|_{\infty, Q_{i,2}^{\varepsilon_0}} dx.$$

where $\delta^{i,r} = \sum_{k=1}^{r} \gamma^{i,k}$. Applying the Hölder inequality with power $p = 2$ to the right–hand side part of the last relation, we obtain

$$\left| \frac{1}{\varepsilon_2} \Delta_\varepsilon I_1^{(2)}(\tau) \right| \leq \sum_{i=1}^{i_1} \sum_{r=1}^{r_0(i)} 2\|M_t \mathfrak{F}_{i,r}(\cdot, t^i)\|_{2, Q_{i,1}^{\varepsilon_0}} \left[\int_{Q_{i,1}^{\varepsilon_0}} \|\Delta_\varepsilon \mathfrak{p}_1(x, \cdot, \tau)\|_{\infty, Q_{i,2}^{\varepsilon_0}}^2 dx \right]^{\frac{1}{2}}.$$

Applying Lemma 3.12 (with $\mathcal{A}_1 \equiv Q_{i,1}^{\varepsilon_0}$, $\mathcal{A}_2 \equiv Q_{i,2}^{\varepsilon_0}$) to the expression

$$\left[\int_{Q_{i,1}^{\varepsilon_0}} \|\Delta_\varepsilon \mathfrak{p}_1(x, \cdot, \tau)\|_{\infty, Q_{i,2}^{\varepsilon_0}}^2 dx \right]^{\frac{1}{2}}, \text{ we conclude that there exists a constant } C >$$

0 such that

$$\left[\int_{Q_{i,1}^{\varepsilon_0}} \|\Delta_\varepsilon \mathfrak{p}_1(x,\cdot,\tau)\|_{\infty,Q_{i,2}^{\varepsilon_0}}^2 \, dx \right]^{\frac{1}{2}} \le C \|\Delta_\varepsilon \mathfrak{p}_1(\cdot,\cdot,\tau)\|_{2,Q_T}^{(1)}.$$

Therefore,

$$\left| \frac{1}{\varepsilon_2} \Delta_\varepsilon I_1^{(2)}(\tau) \right| \le C_1 \sup_{\xi \in [0,T]} \|\Delta_\varepsilon \mathfrak{p}_1(\cdot,\cdot,\xi)\|_{2,Q_T}^{(1)} \sum_{i=1}^{i_1} \sum_{r=1}^{r_0(i)} \|M_t \mathfrak{F}_{i,r}(\cdot,t^i)\|_{2,R^n}.$$

Hence, according to Lemma 3.9 and a definition of variation π_ε,

$$\lim_{\varepsilon_1 \to +0} \frac{1}{\varepsilon_1^n} \lim_{\varepsilon_2 \to +0} \frac{\Delta_\varepsilon I_1^{(2)}(\tau)}{\varepsilon_2} = 0. \tag{3.30}$$

Using an introduced notation, let us rewrite an expression for $\Delta_\varepsilon I_1^{(3)}(\tau)$:

$$\frac{1}{\varepsilon_2} \Delta_\varepsilon I_2^{(3)}(\tau) \equiv \sum_{i=1}^{i_1} \sum_{r=1}^{r_0(i)} \int_{Q_{i,r,1}^{\varepsilon_1}} \left[\frac{1}{\varepsilon_2} \int_{Q_{i,r,2}^{\varepsilon_2}} \mathfrak{F}_{i,r}(x,t) \mathfrak{p}_1[\pi](x,t,\tau) dt \right] dx.$$

Using a maximal function definition, the classic Lebesgue dominated convergence theorem, and the (l,m)–Lebesgue points definition, we obtain that

$$\lim_{\varepsilon_1 \to +0} \frac{1}{\varepsilon_1^n} \lim_{\varepsilon_2 \to +0} \frac{1}{\varepsilon_2} \Delta_\varepsilon I_1^{(3)}(\tau) =$$

$$= -\sum_{i=1}^{i_1} \sum_{r=1}^{r_0(i)} \gamma^{i,r} \mathfrak{p}_1[\pi](x^i,t^i,\tau) \Delta_u a(x^i,t^i,z[\pi](x^i,t^i); u^{i,r}, u(x^i,t^i)).$$

Combining this equality, (3.26), (3.27) and (3.30), we get second assertion of the lemma. This completes the proof of the lemma.

4. The Proof of the Maximum Pinciple

Let us give only the scheme of the proof of theorem 2.1. Suppose $\pi^k \equiv (u^k, v^k, w^k)$, $k = 1, 2, \ldots$, is a m.a.s. in problem (P_q). Consider the problem

$$J(\pi) \to \inf, \quad \pi \in \mathcal{D}, \tag{4.1}$$

where $J(\pi) \equiv \max\{I_0(\pi) - \beta(q), \; I_1(\pi)(\tau) - q(\tau), \; \tau \in X\}$. Obviously, the sequence π^k, $k = 1, 2, \ldots$, is also a minimizing sequence in problem (4.1), and the value of problem (4.1) is equal to zero. Let $\hat{X}^k \equiv \{\tau^{k,j} : j = 1, \ldots, l_k\} \subset X$ be a $1/k$-net in X, $\hat{X}^k \subseteq \hat{X}^{k+1}$, $k = 1, 2, \ldots$. Consider the family of auxiliary problems $J_k(\pi) \to \inf$, $\pi \in \mathcal{D}$, where $J_k(\pi) \equiv \max\{I_0(\pi) - \beta(q); \; I_1(\pi)(\tau) - q(\tau), \; \tau \in \hat{X}^k\}$. By virtue of Lemma 3.4, a functional $J_k(\cdot)$ is continuous and bounded on $\mathcal{D}$. Using a precompactness of the family $\{I_1(\pi) : \pi \in \mathcal{D}\} \subset C(X)$ in the space $C(X)$ (see Lemma 3.4), it is not hard to show that $\lim\limits_{k\to\infty} \inf\limits_{\pi \in \mathcal{D}} J_k(\pi) = \inf\limits_{\pi \in \mathcal{D}} J(\pi) = \lim\limits_{k\to\infty} J(\pi^k) = 0$. It follows that there exists a sequence $\varkappa^k \geq 0$, $k = 1, 2, \ldots$, $\varkappa^k \to 0$, $k \to \infty$, such that $J_k(\pi^k) \leq \inf\limits_{\pi \in \mathcal{D}} J_k(\pi) + \varkappa^k$. Hence, according to the Ekeland variational principle [33], there exists a sequence $\bar{\pi}^k \equiv (\bar{u}^k, \bar{v}^k, \bar{w}^k) \in \mathcal{D}$, $k = 1, 2, \ldots$, such that for any $k = 1, 2, \ldots$, $\bar{\pi}^k$ is a solution to the problem

$$J_k(\pi) + \sqrt{\varkappa^k} d(\bar{\pi}^k, \pi) \to \min, \quad \pi \in \mathcal{D}, \qquad (4.2)$$

and

$$d(\pi^k, \bar{\pi}^k) \leq \sqrt{\varkappa^k}, \quad J_k(\bar{\pi}^k) \leq J_k(\pi^k). \qquad (4.3)$$

Suppose $\pi \in \mathcal{D}$, $I \in R$. By definition, put

$$\hat{I}_j(\pi) = \begin{cases} I_0(\pi), & j = 0; \\ I_1(\pi)(\tau^{k,j}), & j = \overline{1, l_k}; \end{cases} \qquad F_j(I) = \begin{cases} I - \beta(q), & j = 0; \\ I - q(\tau^{k,j}), & j = \overline{1, l_k}. \end{cases}$$

Then, obviously, $J_k(\pi) \equiv \max\limits_{s=0,l_k} F_j(\hat{I}_j(\pi))$. Let $\Gamma_k \equiv \{j = \overline{0, l_k} : J_k(\bar{\pi}^k) = F_j(\hat{I}_j(\bar{\pi}^k))\}$. Then

$$J_k(\bar{\pi}^k) = F_j(\hat{I}_j(\bar{\pi}^k)), \; j \in \Gamma_k; \quad J_k(\bar{\pi}^k) > F_j(\hat{I}_j(\bar{\pi}^k)), \; j \notin \Gamma_k.$$

Let $\bar{k}$ be a number of elements in Γ_k. Introduce first variations vector $\delta\hat{I}(\bar{\pi}^k; \mathrm{m}) \equiv (\delta\hat{I}_0(\bar{\pi}^k; \mathrm{m}), \delta\hat{I}_1(\bar{\pi}^k; \mathrm{m}), \ldots, \delta\hat{I}_{l_k}(\bar{\pi}^k; \mathrm{m})) \in R^{l_k+1}$. By $\mathcal{K}(\bar{\pi}^k)$ denote the set of all first variations vectors. Using standard methods of optimal control, it can be shown that $\mathcal{K}(\bar{\pi}^k)$ is convex. Let us project $\mathcal{K}(\bar{\pi}^k)$ on the subspace of R^{l_k+1} spanned vectors e_j, $j \in \Gamma_k$ (e_j, $j = \overline{0, l_k}$, is the standard basis of R^{l_k+1}), and let us denote the projection by $\mathcal{K}_{\bar{k}}(\bar{\pi}^k)$. Consider the set $\mathcal{K}_{\bar{k}}^- \equiv \{\sum\limits_{j \in \Gamma_k} x_j e_j : x_j \leq -2L\sqrt{\varkappa^k}, \; j \in \Gamma_k\}$, where the constant

$L \equiv 1 + measV + 2A$ is defined in Lemma 3.11. Let us show that the following lemma holds.

Lemma 4.1 $\mathcal{K}_{\bar{k}}^- \cap \mathcal{K}_{\bar{k}}(\bar{\pi}^k) = \emptyset$.

Proof. Assume the converse. Then there exists $\mathfrak{m} \in \mathfrak{M}$ such that $\delta \hat{I}_j(\bar{\pi}^k; \mathfrak{m}) \leq -2L\sqrt{\varkappa^k}$, $j \in \Gamma_k$. In view of first variations' form of functionals in Lemma 3.13, we get $F_j(\hat{I}_j(\bar{\pi}_\varepsilon^k)) - F_j(\hat{I}_j(\bar{\pi}^k)) = \hat{I}_j(\bar{\pi}_\varepsilon^k) - \hat{I}_j(\bar{\pi}^k) = \varepsilon_1^n \varepsilon_2 \delta \hat{I}_j(\bar{\pi}^k; \mathfrak{m}) + \varepsilon_1^n \varepsilon_2 \omega_1(\varepsilon_1) + \varepsilon_2 \omega_2(\varepsilon_1, \varepsilon_2)$, where ω_1 and ω_2 are such that $\lim_{\varepsilon_2 \to 0} \omega_2(\varepsilon_1, \varepsilon_2) = \lim_{\varepsilon_1 \to 0} \omega_1(\varepsilon_1) = 0$. Hence,

$$F_j(\hat{I}_j(\bar{\pi}_\varepsilon^k)) - F_j(\hat{I}_j(\bar{\pi}^k)) = \hat{I}_j(\bar{\pi}_\varepsilon^k) - \hat{I}_j(\bar{\pi}^k) = \varepsilon_1^n \varepsilon_2 \delta \hat{I}_j(\bar{\pi}^k; \mathfrak{m}) + \varepsilon_1^n \varepsilon_2 \omega_1(\varepsilon_1) +$$

$$+ \varepsilon_2 \omega_2(\varepsilon_1, \varepsilon_2) \leq -\frac{19}{10} L \varepsilon_1^n \varepsilon_2 \sqrt{\varkappa^k} + \varepsilon_1^n \varepsilon_2 \omega_1(\varepsilon_1) + [-\frac{19}{10} L \varepsilon_1^n \sqrt{\varkappa^k} + \omega_2(\varepsilon_1, \varepsilon_2)] \varepsilon_2 <$$

$$< -\frac{19}{10} L \varepsilon_1^n \varepsilon_2 \sqrt{\varkappa^k} + \varepsilon_1 \varepsilon_2^n \omega_1(\varepsilon_1) = -\frac{9}{5} L \varepsilon_1^n \varepsilon_2 \sqrt{\varkappa^k} + [-\frac{1}{10} L \sqrt{\varkappa^k} + \omega_1(\varepsilon_1)] \varepsilon_1^n \varepsilon_2 <$$

$$< -\frac{9}{5} L \varepsilon_1^n \varepsilon_2 \sqrt{\varkappa^k}$$

for all enough small ε. Therefore, $F_j(\hat{I}_j(\bar{\pi}_\varepsilon^k)) \leq F_j(\hat{I}_j(\bar{\pi}^k)) - \frac{9}{5} L \varepsilon_1^n \varepsilon_2 \sqrt{\varkappa^k}$, $j \in \Gamma_k$. According to Lemma 3.11, it follows that

$$F_j(\hat{I}_j(\bar{\pi}_\varepsilon^k)) + \sqrt{\varkappa^k} d(\bar{\pi}_\varepsilon^k, \bar{\pi}^k) \leq F_j(\hat{I}_j(\bar{\pi}^k)) - \frac{9}{5} L \varepsilon_1^n \varepsilon_2 \sqrt{\varkappa^k} +$$

$$+ L \varepsilon_1^n \varepsilon_2 \sqrt{\varkappa^k} + \sqrt{\varkappa^k} d(\bar{\pi}^k, \bar{\pi}^k) = J_k(\bar{\pi}^k) + \sqrt{\varkappa^k} d(\bar{\pi}^k, \bar{\pi}^k) - \frac{4}{5} L \varepsilon_1^n \varepsilon_2 \sqrt{\varkappa^k}.$$

Thus, $J_k(\bar{\pi}_\varepsilon^k) + \sqrt{\varkappa^k} d(\bar{\pi}_\varepsilon^k, \bar{\pi}^k) < J_k(\bar{\pi}^k) + \sqrt{\varkappa^k} d(\bar{\pi}^k, \bar{\pi}^k)$. But this contradicts to an optimality of $\bar{\pi}^k$ in problem (4.2). The lemma is proved.

Since $\mathcal{K}_{\bar{k}}^- \cap \mathcal{K}_{\bar{k}}(\bar{\pi}^k) = \emptyset$, these sets are separable; i.e., there exists a vector $\lambda^{\bar{k}} \in R^{\bar{k}}$, $\lambda_j^{\bar{k}} \geq 0$, $j \in \Gamma_k$, $\sum_{j \in \Gamma_k} \lambda_j^{\bar{k}} = 1$, such that

$$\sum_{j \in \Gamma_k} \lambda_j^{\bar{k}} \delta \hat{I}_j(\bar{\pi}^k; \mathfrak{m}) \geq \sum_{j \in \Gamma_k} \lambda_j^{\bar{k}} x_j \quad \forall \mathfrak{m} \in \mathfrak{M} \; \forall x = \sum_{j \in \Gamma_k} x_j e_j \in \mathcal{K}_{\bar{k}}^-.$$

Putting $x_j = -2L\sqrt{\varkappa^k}$, $j \in \Gamma_k$, in the last inequality, and completing vectors

$\lambda^{\bar{k}} \in R^{\bar{k}}$ to vectors $\lambda^k \in R^{l_k+1}$ by zeros, we conclude that

$$\lambda_j^k \geq 0, \quad j = 0, \ldots, l_k; \quad \lambda_0^k + \sum_{j=1}^{l_k} \lambda_j^k = 1; \tag{4.4}$$

$$\lambda_j^k (J_k(\bar{\pi}^k) - (I_1(\bar{\pi}^k)(\tau^{k,j}) - q(\tau^{k,j}))) = 0, \quad j = 1, \ldots, l_k; \tag{4.5}$$

$$\lambda_0^k \delta I_0(\bar{\pi}^k; \mathfrak{m}) + \sum_{j=1}^{l_k} \delta I_1(\bar{\pi}^k, \tau^{k,j}; \mathfrak{m}) \geq -2L\sqrt{\varkappa^k}, \quad \forall \mathfrak{m} \in \mathfrak{M}. \tag{4.6}$$

In view of first variations expressions (see Lemma 3.13), from the last inequality it follows that

$$H(x, t, z[\bar{\pi}^k](x, t), u, \lambda_0^k \mathfrak{p}_0[\bar{\pi}^k](x, t) + \sum_{j=1}^{l_k} \lambda_j^k \mathfrak{p}_1[\bar{\pi}^k](x, t, \tau^{k,j})) - \tag{4.7}$$

$$-H(x, t, z[\bar{\pi}^k](x, t), \bar{u}^k(x, t), \lambda_0^k \mathfrak{p}_0[\bar{\pi}^k](x, t) + \sum_{j=1}^{l_k} \lambda_j^k \mathfrak{p}_1[\bar{\pi}^k](x, t, \tau^{k,j}))$$

$$\leq 2L\sqrt{\varkappa^k} \forall\ u \in U \text{ for a.e. } (x, t) \in Q_T;$$

$$\int_\Omega [\lambda_0^k \mathfrak{p}_0[\bar{\pi}^k](x, 0) + \sum_{j=1}^{l_k} \lambda_j^k \mathfrak{p}_1[\bar{\pi}^k](x, 0, \tau^{k,j})](\bar{v}^k(x) - \tilde{v}(x))\, dx + \tag{4.8}$$

$$+ \sum_{j=1}^{l_k} \lambda_j^k \int_\Omega \nabla_v \Phi(x, \tau^{k,j}, z[\bar{\pi}^k](x, \tau^{k,j}), \bar{v}^k)(\bar{v}^k(x) - \tilde{v}(x)) dx \leq$$

$$\leq 2\sqrt{\varkappa^k}\ \forall \tilde{v} \in \mathcal{D}_2;$$

$$\int_{S_T} [\lambda_0^k \mathfrak{p}_0[\bar{\pi}^k](s, t) + \sum_{j=1}^{l_k} \lambda_j^k \mathfrak{p}_1[\bar{\pi}^k](s, t, \tau^{k,j})](\bar{w}^k - \tilde{w})\, ds dt \leq 2\sqrt{\varkappa^k} \tag{4.9}$$

$$\forall\ \tilde{w} \in \mathcal{D}_3.$$

By definition, put $\mu^k \equiv \sum_{j=1}^{l_k} \lambda_j^k \delta_{\tau^{k,j}}$, where δ_τ is a Radon δ–measure concentrated in the point τ. Then $\sum_{j=1}^{l_k} \lambda_j^k \mathfrak{p}_1[\bar{\pi}^k](x, t, \tau^{k,j}) = \int_X \mathfrak{p}_1[\bar{\pi}^k](x, t, \tau)\mu^k(d\tau)$. According to Theorem 3.1, $\int_X \mathfrak{p}_1[\bar{\pi}^k](x, t, \tau)\mu^k(d\tau) \equiv \mathfrak{f}[\mathfrak{a}[\bar{\pi}^k], \mathfrak{g}_1[\bar{\pi}^k], \mu^k](x, t)$.

Hence, relations (4.4), (4.5), (4.7)–(4.9) can be rewritten in the form

$$\lambda_0^k \geq 0, \quad \lambda_0^k + \|\mu^k\| = 1; \tag{4.10}$$

$$\|\mu^k\| J_k(\bar{\pi}^k) - \int_X [I_1(\bar{\pi}^k)(\tau) - q(\tau)]\mu^k(d\tau) = 0; \tag{4.11}$$

$$H(x, t, z[\bar{\pi}^k](x, t), u, \eta[\bar{\pi}^k, \lambda_0^k, \mu^k](x, t)) - \tag{4.12}$$
$$-H(x, t, z[\bar{\pi}^k](x, t), \bar{u}^k(x, t), \eta[\bar{\pi}^k, \lambda_0^k, \mu^k](x, t)) \leq 2L\sqrt{\varkappa^k}$$
$$\forall \, u \in U \text{ for a.e. } (x, t) \in Q_T;$$

$$\int_\Omega \eta[\bar{\pi}^k, \lambda_0^k, \mu^k](x, 0)(\bar{v}^k(x) - \tilde{v}(x)) \, dx + \tag{4.13}$$

$$+ \int_X \mu^k(d\tau) \int_\Omega \nabla_v \Phi(x, \tau, z[\bar{\pi}^k](x, \tau), \bar{v}^k(x))(\bar{v}^k(x) - \tilde{v}(x)) \, dx \leq 2L\sqrt{\varkappa^k}$$
$$\forall \, \tilde{v} \in \mathcal{D}_2;$$

$$\int_{S_T} \eta[\bar{\pi}^k, \lambda_0^k, \mu^k](s, t)(\bar{w}^k - \tilde{w}) \, ds dt \leq 2L\sqrt{\varkappa^k} \; \forall \, \tilde{w} \in \mathcal{D}_3. \tag{4.14}$$

Using assumptions on source data of problem (P_q), Lemma 3.2, Lemma 3.3, Lemma 3.9, the first inequality in (4.3), from relations (4.12), (4.13), and (4.14) we obtain relations (2.3), (2.4), and (2.5) respectively. From the second inequality in (4.3), relation (4.11), and an uniform continuity of I_1 on $\mathcal{D}$ (see Lemma 3.4), it follows that the nondegenerate condition holds, and the measure μ^k is concentrated on the set X_k. This completes the proof of Theorem 2.1.

5. Approximation of Problem (P_q) by Problems with Finite Numbers of Constraints

To discuss proofs of other main results, let us approximate problem (P_q) by problems with finite numbers of functional constraints. Let $\hat{X}^k \equiv \{\tau^{k,j} : j = 1, \ldots, l_k\} \subset X$, $\hat{X}^k \subset \hat{X}^{k+1}$, be a $1/k$–net in X.

Consider the following sequence of families of optimization problems depending on a finite–dimensional vector parameter $q^k \equiv (q_1^k, \ldots, q_{l_k}^k) \in R^{l_k}$ that approximate the original family (P_q):

$$I_0(\pi) \to \inf, \; I^k(\pi) \in \mathcal{M}^k + q^k, \; \pi \in \mathcal{D}, \qquad q^k \in R^{l_k} \text{ is a parameter}, \quad (P_{q^k}^k)$$

where $\mathcal{M}^k \equiv \{y \in R^{l_k} : y_i \le 0, \ i = \overline{1, l_k}\}$, $I^k(\pi) \equiv (I_1^k(\pi), \ldots, I_{l_k}^k(\pi))$, $I_i^k(\pi) \equiv I_1(\pi)(\tau^{k,i})$, $\tau^{k,i} \in \hat{X}^k$. As in the case of problem (P_q), the value function $\beta_k : R^{l_k} \to R \cup \{+\infty\}$ of problem $(P_{q^k}^k)$ is defined by

$$\beta_k(q^k) \equiv \lim_{\varepsilon \to +0} \beta_{k,\varepsilon}(q^k), \quad \beta_{k,\varepsilon}(q^k) \equiv \begin{cases} \inf_{\pi \in \mathcal{D}_{q^k}^{k,\varepsilon}} I_0(\pi), \ \mathcal{D}_{q^k}^{k,\varepsilon} \ne \emptyset; \\ +\infty, \ \mathcal{D}_{q^k}^{k,\varepsilon} = \emptyset, \end{cases}$$

where $\mathcal{D}_{q^k}^{k,\varepsilon} \equiv \{\pi \in \mathcal{D} : I_j^k(\pi) \le q_j^k + \varepsilon, \ j = 1, \ldots, l_k\}$. The following approximation lemma is valid.

Lemma 5.1 *Suppose $\beta(q) < +\infty$, $q \in C(X)$; then the sequence of vectors $\bar{q}^k \in R^{l_k}$, $\bar{q}^k \equiv (\bar{q}_1^k, \ldots, \bar{q}_{l_k}^k)$, $\bar{q}_i^k = q(\tau^{k,i})$, $i = 1, \ldots, l_k$, $k = 1, 2, \ldots$, satisfies the relation $\beta_k(\bar{q}^k) \to \beta(q)$ as $k \to \infty$.*

Proof. Let π^r, $r = 1, 2, \ldots$, be a sequence of triples such that $\pi^r \in \mathcal{D}_q^{\varepsilon^r}$ and $I_0(\pi^r) \to \beta(q)$, $\varepsilon^r \ge 0$, $\varepsilon^r \to 0$, $r \to \infty$. According to lemma 3.4, the functions of the family $I_1(\pi^r)$, $r = 1, 2, \ldots$, are uniformly bounded and equicontinuous on X. Therefore, we may assume without loss of generality that

$$|I_1(\pi^r) - \hat{q}|_X^{(0)} \to \infty, \ r \to \infty; \ \hat{q}(\tau) \le q(\tau), \ \tau \in X;$$
$$\max\{0; I_j^k(\pi^r) - \bar{q}_j^k, \ j = \overline{1, l_k}\} \to 0, \ r \to \infty.$$

Since, additionally, $\hat{X}^k \subseteq \hat{X}^{k+1}$, $k = 1, 2, \ldots$, one can select a subsequence r_k, $k = 1, 2, \ldots$, of a sequence $r = 1, 2, \ldots$, such that $I_0(\pi^{r_k}) \ge \beta_k(\bar{q}^k) - \gamma^k$, $\gamma^k \ge 0$, and $\gamma^k \to 0$ as $k \to \infty$. Hence, it follows that $\limsup_{k \to \infty} \beta_k(\bar{q}^k) \le \beta(q)$. Let us show that $\liminf_{k \to \infty} \beta_k(\bar{q}^k) \ge \beta(q)$, which, clearly, proves the assertion. Suppose that this is not true and, without loss of generality, $\lim_{k \to \infty} \beta_k(\bar{q}^k) = \alpha < \beta(q)$. Then, there exists a sequence π^r, $r = 1, 2, \ldots$, such that $I_0(\pi^r) \to \alpha$ and $\max\{0; I_j^k(\pi^r) - \bar{q}_j^k, \ j = \overline{1, l_r}\} \to 0$ as $r \to \infty$, which contradicts the assumption that $\beta(q)$ is a value of problem (P_q).

In what follows, we need following definitions and some results from [30] – [32]. Let $\Theta \subset R^s$ be a nonempty set, $\varepsilon \ge 0$, $x \in \bar{\Theta}$. The nonempty set

$$\hat{N}_\varepsilon(x; \Theta) \equiv \{x^* \in R^s : \limsup_{u \xrightarrow{\Theta} x} \frac{\langle x^*, u - x \rangle}{|u - x|} \le \varepsilon\}$$

is said to be the set of Frechet ε-normals to the set Θ at the point x. Here, the notation $u \xrightarrow{\Theta} x$ means that $u \to x$ for $u \in \Theta$. In particular, $\hat{N}_0(x; \Theta)$ is called the cone of Frechet normals to the set Θ at the point x and is denoted by $\hat{N}(x; \Theta)$. The normal (basic, limit) cone at a point $\bar{x} \in \Theta$ is defined as
$$N(\hat{x}; \Theta) \equiv \limsup_{x \xrightarrow{\Theta} \hat{x}, \varepsilon \downarrow 0} \hat{N}_\varepsilon(x; \Theta).$$

It can be shown (see [30]–[32] for details) that $N(\hat{x}; \Theta) \equiv \limsup_{x \xrightarrow{\Theta} \hat{x}} \hat{N}(x; \Theta)$.

For a function $f: R^s \to R \cup \{+\infty\}$ lower semicontinuous and $\bar{x} \in \operatorname{dom} f$, the Frechet subdifferential $\hat{\partial}f(\bar{x})$ of the function f at the point $\bar{x} \in \operatorname{dom} f$ is defined as

$$\hat{\partial}f(\bar{x}) \equiv \left\{ x^* \in R^s : \liminf_{x \to \bar{x}} \frac{f(x) - f(\bar{x}) - \langle x^*, x - \bar{x} \rangle}{|x - \bar{x}|} \geq 0 \right\},$$

or, equivalently, as

$$\hat{\partial}f(\bar{x}) \equiv \left\{ x^* \in R^s : (x^*, -1) \in \hat{N}((\bar{x}, f(\bar{x})); \operatorname{epi} f) \right\}.$$

For any $\bar{x} \in \operatorname{dom} f$, the sets

$$\partial f(\bar{x}) \equiv \{x^* \in R^s : (x^*, -1) \in N((\bar{x}, f(\bar{x})); \operatorname{epi} f)\},$$
$$\partial^\infty f(\bar{x}) \equiv \{x^* \in R^s : (x^*, 0) \in N((\bar{x}, f(\bar{x})); \operatorname{epi} f)\},$$

are called, respectively, the subdifferential and singular subdifferential of the function f at the point $\bar{x}$ in the sense of [30]–[32]. If the function f is lower semicontinuous, then the following relations hold:

$$\partial f(\bar{x}) = \limsup_{x \xrightarrow{f} \bar{x}} \hat{\partial}f(x), \quad \partial^\infty f(\bar{x}) = \limsup_{x \xrightarrow{f} \bar{x}; \bar{\varepsilon} \downarrow 0} \bar{\varepsilon}\hat{\partial}f(x), \qquad (5.1)$$

where $x \xrightarrow{f} \bar{x}$ means that $x \to \bar{x}$, $f(x) \to f(\bar{x})$. If f is Lipschitz in a neighborhood of x, then $\partial^\infty f(\bar{x}) = \{0\}$.

The following result is very important (see [30]–[32]).

Lemma 5.2 *Let $\Theta \subset R^s$ be a nonempty closed set. Then, the set of points $\{x \in \Theta : \hat{N}(x; \Theta) \neq \{0\}\}$ (i.e., the set of all boundary points of Θ for which there exists nonzero Frechet normal) is everywhere dense in the set of all boundary points of Θ. In addition, for any functions $f: R^s \to R \cup \{\pm\infty\}$ that is lower* *semicontinuous, the set $\{x \in \operatorname{dom} f : \hat{\partial} f(x) \neq \emptyset\}$ is everywhere dense in* $\operatorname{dom} f$.

From the definition of the Frechet subdifferential of a function f at a point x, we immediately obtain the following lemma.

Lemma 5.3 *Let $f \colon R^s \to R \cup \{\pm\infty\}$ be a lower semicontinuous function and* $x \in \operatorname{dom} f$. *If $(x^*, -\eta) \in \hat{N}((x, f(x)); epi\, f)$, $\eta > 0$, then, for any $\varepsilon > 0$ there exists a neighborhood S_ε of x such that $\eta f(x') - \eta f(x) - \langle x^*, x' - x \rangle + \varepsilon |x' - x| \geq 0 \ \forall\, x' \in S_\varepsilon$.*

Now, we prove the following lemma on the relationship between the Lagrange multipliers and Frechet normals in the sense of [30]–[32].

Lemma 5.4 *Let $\beta_k(q^k) < \infty$ and $(\zeta^k, -\varkappa^k) \in \hat{N}((q^k, \beta_k(q^k)); epi\beta_k)$ be an arbitrary vector. Then, there exist a sequence of nonnegative numbers $\gamma^i \to 0$, $i \to \infty$, a sequence of controls $\pi^i \in \mathcal{D}_{q^k}^{k,\gamma^i}$, $i = 1, 2, \ldots$, and a bounded sequence of the Lagrange multipliers $\lambda^i \equiv (\lambda_0^i, \ldots, \lambda_{l_k}^i) \in R^{l_k}$, $i = 1, 2, \ldots$, such that*

$$|\lambda^i| \neq 0, \ \lambda_j^i \geq 0, \ j = \overline{0, l_k}; \ \lambda_j^i(I_j^k(\pi^i) - q_j^k) \geq -\gamma^i, \ j = \overline{1, l_k}; \quad (5.2)$$

$$\int_{Q_T} [\max_{u' \in U} H(x, t, z[\pi^i](x,t), u', \eta[\pi^i, \lambda_0^i, \mu^i](x,t)) - \quad (5.3)$$

$$- H(x, t, z[\pi^i](x,t), u^i(x,t), \eta[\pi^i, \lambda_0^i, \mu^i](x,t))]\, dx dt \leq \gamma^i;$$

$$\max_{v \in \mathcal{D}_2} \left\{ \int_\Omega \eta[\pi^i, \lambda_0^i, \mu^i](x, 0)(v^i(x) - v(x))\, dx + \right. \quad (5.4)$$

$$\left. + \int_X \mu^i(d\tau) \int_\Omega \nabla_v \Phi(x, \tau, z[\pi^i](x,\tau), v^i(x))(v^i(x) - v(x))\, dx \right\} \leq \gamma^i,$$

$$\max_{w \in \mathcal{D}_3} \int_{S_T} \eta[\pi^i, \lambda_0^i, \mu^i](s,t)(w^i - w)\, ds dt \leq \gamma^i, \ i = 1, 2, \ldots; \quad (5.5)$$

$$\zeta^k + \sum_{j=1}^{l_k} \lambda_j e^j = 0; \quad (5.6)$$

where $\mu^i \equiv \sum_{j=1}^{l_k} \lambda_j^i \delta_{\tau^{k,j}}$, $\lambda \equiv (\varkappa^k, \lambda_1, \ldots, \lambda_{l_k}) \neq 0$ is a limit point of the sequence of vectors λ^i, $i = 1, 2, \ldots$, δ_τ is Radon δ–measure concentrated at the point $\tau \in X$, and $e^j \equiv (\underbrace{0, \ldots, 0}_{j-1}, 1, 0, \ldots, 0) \in R^{l_k}$, $j = \overline{1, l_k}$.

Proof. First, we prove the lemma for $\varkappa^k > 0$. Let $q^{k,i} \in R^{l_k}$, $\varkappa^{k,i} \in R$, and $\zeta^{k,i} \in R^{l_k}$, $i = 1, 2, \ldots$, be sequences such that

$$(\zeta^{k,i}, -\varkappa^{k,i}) \in \hat{N}((q^{k,i}, \beta_k(q^{k,i})); \text{ epi } \beta_k), \tag{5.7}$$

$$q^{k,i} \to q^k, \quad \beta_k(q^{k,i}) \to \beta_k(q^k), \quad (\zeta^{k,i}, -\varkappa^{k,i}) \to (\zeta^k, -\varkappa^k), \quad i \to \infty.$$

The existence of sequences $q^{k,i}$, $\varkappa^{k,i}$, and $\zeta^{k,i}$, $i = 1, 2, \ldots$, with the required properties follows from the first formula in (5.1) for $\partial \beta_k(q^k)$. ¿From (5.7) and lemma 5.3, it follows that, for any $\rho > 0$, there exists a ball $\bar{S}^{l_k}_{\omega(\rho)}(q^{k,i})^\dagger$ such that the point $q^{k,i}$ is a minimizier in the problem

$$\varkappa^{k,i} \beta_k(q^{k,i}) - \langle \zeta^{k,i}, q' \rangle + \rho |q' - q^{k,i}| \to \inf, \quad q' \in \bar{S}^{l_k}_{\omega(\rho)}(q^{k,i}).$$

Let us prove that, if a sequence $\pi^{i,\ell}$, $\ell = 1, 2, \ldots$, is a m.a.s in problem $(P^k_{q^{k,i}})$, then the sequence of pairs $(\pi^{i,\ell}, q^{k,i})$, $\ell = 1, 2, \ldots$, is a m.a.s. in the problem

$$I^i(\pi, q') \to \inf, \quad I^k_1(\pi) - q' \in \mathcal{M}^k, \quad (\pi, q') \in \hat{\mathcal{D}} \equiv \mathcal{D} \times \bar{S}^{l_k}_{\omega(\rho)}(q^{k,i}), \tag{5.8}$$

where $I^i(\pi, q') \equiv \varkappa^{k,i} I_0(\pi) - \langle \zeta^{k,i}, q' \rangle + \rho |q' - q^{k,i}|$. First of all, we equip $\hat{\mathcal{D}}$ with the metric $\hat{d}((\pi^1, q^1), (\pi^2, q^2)) \equiv d(\pi^1, \pi^2) + |q^1 - q^2|$ and introduce the notation $\hat{\mathcal{D}}^\alpha \equiv \{(\pi, q') \in \hat{\mathcal{D}} : I^k_j(\pi) - q'_j \leq \alpha, \ j = \overline{1, l_k}\}$, $\hat{\beta}^i_k \equiv \lim_{\alpha \to 0} \hat{\beta}^{i,\alpha}_k$,
$\hat{\beta}^{i,\alpha}_k \equiv \inf_{(\pi, q') \in \hat{\mathcal{D}}^\alpha} I^i(\pi, q')$.

In accordance with the general definition [21], the sequence $(\bar{\pi}^\ell, q'^{\,\ell}) \in \hat{\mathcal{D}}$, $\ell = 1, 2, \ldots$, is a m.a.s. in problem (5.8) if there exist number sequences $\bar{\gamma}^\ell$, $\bar{\varepsilon}^\ell \geq 0$, $\bar{\gamma}^\ell$, $\bar{\varepsilon}^\ell \to 0$, $\ell \to \infty$, such that

$$I^i(\bar{\pi}^\ell, q'^{\,\ell}) \leq \hat{\beta}^i_k + \bar{\gamma}^\ell, \quad (\bar{\pi}^\ell, q'^{\,\ell}) \in \hat{\mathcal{D}}^{\bar{\varepsilon}^\ell}, \quad \ell = 1, 2, \ldots. \tag{5.9}$$

Since $\pi^{i,\ell}$, $\ell = 1, 2, \ldots$, is a m.a.s. in problem $(P^k_{q^{k,i}})$, there exists nonnegative number sequences $\varepsilon^{i,\ell}$, $\gamma^{i,\ell} \to 0$, $\ell \to \infty$, such that $\pi^{i,\ell} \in \mathcal{D}^{k,\varepsilon^{i,\ell}}_{q^{k,i}}$ and $I_0(\pi^{i,\ell}) \leq \beta_k(q^{k,i}) + \gamma^{i,\ell}$. Therefore,

$$I^i(\pi^{i,\ell}, q^{k,i}) = \varkappa^{k,i} I_0(\pi^{i,\ell}) - \langle \zeta^{k,i}, q^{k,i} \rangle \leq \tag{5.10}$$
$$\leq \varkappa^{k,i} \beta_k(q^{k,i}) - \langle \zeta^{k,i}, q^{k,i} \rangle + \varkappa^{k,i} \gamma^{i,\ell}.$$

---

†Here and in what follows, we use the notation $S^n_r(\bar{x}) \equiv \{x \in R^n : |x - \bar{x}| < r\}$.

Since the lower bound $\hat{\beta}_k^i$ in problem (5.8) is equal to $\hat{\beta}_k^i = \varkappa^{k,i}\beta_k(q^{k,i}) - \langle \zeta^i, q^{k,i} \rangle$ (this fact can be proved similarly to equality (5.6) in [38]), estimate (5.10) can be rewritten as

$$I^i(\pi^{i,\mathfrak{k}}, q^{k,i}) \leq \hat{\beta}_k^i + \tilde{\gamma}^{i,\mathfrak{k}}, \qquad (5.11)$$

where $\tilde{\gamma}^{i,\mathfrak{k}} \equiv \varkappa^{k,i}\gamma^{i,\mathfrak{k}}$. Moreover, from the relations $\varepsilon^{i,\mathfrak{k}} \to 0$, $\mathfrak{k} \to \infty$, and $\pi^{i,\mathfrak{k}} \in \mathcal{D}_{q^{k,i}}^{k,\varepsilon^{i,\mathfrak{k}}}$ and the definition of the set $\hat{\mathcal{D}}^\alpha$, it follows that $(\pi^{i,\mathfrak{k}}, q^{k,i}) \in \hat{\mathcal{D}}^{\varepsilon^{i,\mathfrak{k}}}$. The last inclusion, together with estimate (5.11), proves that the sequence of pairs $(\pi^{i,\mathfrak{k}}, q^{k,i})$ is actually a m.a.s. in problem (5.8) and that the sequence $r^{i,\mathfrak{k}} \equiv \hat{\beta}_k^i - \hat{\beta}_k^{i,\varepsilon^{i,\mathfrak{k}}} + \tilde{\gamma}^{i,\mathfrak{k}}$, $i = 1, 2, \ldots$, satisfies the relations

$$I^i(\pi^{i,\mathfrak{k}}, q^{i,k}) \leq \hat{\beta}_k^{i,\varepsilon^{i,\mathfrak{k}}} + r^{i,\mathfrak{k}}, \quad (\pi^{i,\mathfrak{k}}, q^{k,i}) \in \hat{\mathcal{D}}^{\varepsilon^{i,\mathfrak{k}}}.$$

Hence, it follows, with the help of the Ekeland variational principle [33], that there exists a sequence $(\pi^{i,\mathfrak{k},1}, q^{k,i,\mathfrak{k}}) \in \hat{\mathcal{D}}^{\varepsilon^{i,\mathfrak{k}}}$ that is a solution to the problem

$$I^i(\pi, q') + \sqrt{r^{i,\mathfrak{k}}}\hat{d}((\pi, q'), (\pi^{i,\mathfrak{k},1}, q^{k,i,\mathfrak{k}})) \to \inf, \quad (\pi, q') \in \hat{\mathcal{D}}^{\varepsilon^{i,\mathfrak{k}}}, \qquad (5.12)$$

and satisfies the relations

$$\hat{d}((\pi^{i,\mathfrak{k}}, q^{k,i}), (\pi^{i,\mathfrak{k},1}, q^{k,i,\mathfrak{k}})) \leq \sqrt{r^{i,\mathfrak{k}}}, \quad I^i(\pi^{i,\mathfrak{k},1}, q^{k,i,\mathfrak{k}}) \leq I^i(\pi^{i,\mathfrak{k}}, q^{k,i}). \quad (5.13)$$

Let us "smooth out" problem (5.12), i.e., consider the problem

$$I^{i,\alpha}(\pi, q') + \sqrt{r^{i,\mathfrak{k}}}\hat{d}((\pi, q'), (\pi^{i,\mathfrak{k},1}, q^{k,i,\mathfrak{k}})) \to \inf, \quad (\pi, q') \in \hat{\mathcal{D}}^{\varepsilon^{i,\mathfrak{k}}},$$

where $I^{i,\alpha}(\pi, q') \equiv \varkappa^{k,i}I_0(\pi) - \langle \zeta^{k,i}, q' \rangle + \rho|q' - q^{k,i}|^{(\alpha+1)/\alpha}$, $\alpha = 1, 2, \ldots$. Then, $I^{i,\alpha}(\pi, q') - I^i(\pi, q') \to 0$ as $\alpha \to \infty$ uniformly with respect to $(\pi, q') \in \hat{\mathcal{D}}^{\varepsilon^{i,\mathfrak{k}}}$, in view of the boundness of the ball $\bar{S}_{\omega(\rho)}^{lk}(q^{k,i})$. Therefore,

$$\inf_{(\pi,q')\in\hat{\mathcal{D}}^{\varepsilon^{i,\mathfrak{k}}}} [I^{i,\alpha}(\pi, q') + \sqrt{r^{i,\mathfrak{k}}}\hat{d}((\pi, q'), (\pi^{i,\mathfrak{k},1}, q^{k,i,\mathfrak{k}}))] \to I^i(\pi^{i,\mathfrak{k},1}, q^{k,i,\mathfrak{k}}),$$

$$\alpha \to \infty,$$

and, hence, we can again apply the Ekeland variational principle [33]. Thus, there exists a pair $(\pi^{i,\mathfrak{k},1,\alpha}, q^{k,i,\mathfrak{k},\alpha}) \in \hat{\mathcal{D}}^{\varepsilon^{i,\mathfrak{k}}}$ that is a solution to the problem

$$I^{i,\alpha}(\pi, q') + \sqrt{r^{i,\mathfrak{k}}}\hat{d}((\pi, q'), (\pi^{i,\mathfrak{k},1}, q^{k,i,\mathfrak{k}})) + \qquad (5.14)$$

$$+ \sqrt{\mathfrak{b}^\alpha}\hat{d}((\pi, q'), (\pi^{i,\mathfrak{k},1,\alpha}, q^{k,i,\mathfrak{k},\alpha})) \to \inf_{(\pi,q')\in\hat{\mathcal{D}}^{\varepsilon^{i,\mathfrak{k}}}},$$

and satisfies the inequality

$$\hat{d}((\pi^{i,\ell,1}, q^{k,i,\ell}), (\pi^{i,\ell,1,\alpha}, q^{k,i,\ell,\alpha})) \le \sqrt{\mathfrak{b}^\alpha},$$

where $\mathfrak{b}^\alpha > 0$, $\mathfrak{b}^\alpha \to 0$, $\alpha \to \infty$.

Using (5.14) and the definition of the set $\hat{\mathcal{D}}^{\varepsilon^{i,\ell}}$, we conclude that

$$J_\gamma(\pi^{i,\ell,1,\alpha}, q^{k,i,\ell,\alpha}) \le \inf_{\hat{\mathcal{D}}} J_\gamma(\pi, q') + \gamma, \tag{5.15}$$

where

$$J_\gamma(\pi, q') \equiv \max\{I^{i,\alpha}(\pi, q') - I^{i,\alpha}(\pi^{i,\ell,1}, q^{k,i,\ell}) + \gamma +$$
$$+ \sqrt{r^{i,\ell}}\hat{d}((\pi, q'), (\pi^{i,\ell,1}, q^{k,i,\ell})) + \sqrt{\mathfrak{b}^\alpha}\hat{d}((\pi, q'), (\pi^{i,\ell,1,\alpha}, q^{k,i,\ell,\alpha}));$$
$$I_j^k(\pi) - q_j' - \varepsilon^{i,\ell}, \quad j = \overline{1, l_k}\}.$$

Applying again the Ekeland variational principle [33], we find that there exists a pair $(\pi^{i,\ell,1,\alpha,\gamma}, q^{k,i,\ell,\alpha,\gamma}) \in \hat{\mathcal{D}}$ that is a solution to the problem

$$J_\gamma(\pi, q') + \sqrt{\gamma}\hat{d}((\pi, q'), (\pi^{i,\ell,1,\alpha,\gamma}, q^{k,i,\ell,\alpha,\gamma})) \to \inf_{(\pi,q')\in\hat{\mathcal{D}}},$$

and satisfies the inequality

$$\hat{d}((\pi^{i,\ell,1,\alpha}, q^{k,i,\ell,\alpha}), (\pi^{i,\ell,1,\alpha,\gamma}, q^{k,i,\ell,\alpha,\gamma})) \le \sqrt{\gamma}.$$

Applying reasoning similar to that in the proof of the relations (4.7)–(4.9), we get that there exists vector $\lambda^{i,\ell,1,\alpha,\gamma} \equiv (\lambda_0^{i,\ell,1,\alpha,\gamma}, \lambda_1^{i,\ell,1,\alpha,\gamma}, \dots, \lambda_{l_k}^{i,\ell,1,\alpha,\gamma}) \in$

R^{l_k+1} such that

$$\lambda_j^{i,\ell,1,\alpha,\gamma} \geq 0, \ \ j = \overline{0,l_k}, \ \ \lambda_0^{i,\ell,1,\alpha,\gamma} + \sum_{j=1}^{l_k} \lambda_j^{i,\ell,1,\alpha,\gamma} = 1,$$

$$\lambda_j^{i,\ell,1,\alpha,\gamma}[J_\gamma(\pi^{i,\ell,1,\alpha,\gamma}, q^{k,i,\ell,\alpha,\gamma}) - [I_j^k(\pi^{i,\ell,1,\alpha,\gamma}) - q^{k,i,\ell,\alpha,\gamma} - \varepsilon^{i,\ell}]] = 0,$$
$$j = \overline{1,l_k};$$

$$H(x,t,z[\pi^{i,\ell,1,\alpha,\gamma}](x,t),u,\eta[\pi^{i,\ell,1,\alpha,\gamma}, \lambda_0^{i,\ell,1,\alpha,\gamma}\varkappa^{k,i}, \mu^{i,\ell,1,\alpha,\gamma}](x,t)) -$$
$$H(x,t,z[\pi^{i,\ell,1,\alpha,\gamma}](x,t),u^{i,\ell,1,\alpha,\gamma}(x,t),\eta[\pi^{i,\ell,1,\alpha,\gamma}, \lambda_0^{i,\ell,1,\alpha,\gamma}\varkappa^{k,i},$$
$$\mu^{i,\ell,1,\alpha,\gamma}](x,t)) \leq 2L[\sqrt{\gamma} + \sqrt{\mathfrak{b}^\alpha} + \sqrt{r^{i,\ell}}] \ \forall \, u \in U \text{ for a.e. } (x,\,t) \in Q_T;$$

$$\int_\Omega \eta[\pi^{i,\ell,1,\alpha,\gamma}, \lambda_0^{i,\ell,1,\alpha,\gamma}\varkappa^{k,i}, \mu^{i,\ell,1,\alpha,\gamma}](x,0)(v^{i,\ell,1,\alpha,\gamma}(x) - v(x))\,dx +$$

$$\int_X \mu^{i,\ell,1,\alpha,\gamma}(d\tau)\int_\Omega \nabla_v\Phi(x,\tau,z[\pi^{i,\ell,1,\alpha,\gamma}](x,\tau),v^{i,\ell,1,\alpha,\gamma})(v^{i,\ell,1,\alpha,\gamma} - v)\,dx \leq$$
$$\leq 2L[\sqrt{\gamma} + \sqrt{\mathfrak{b}^\alpha} + \sqrt{r^{i,\ell}}] \ \forall \, v \in \mathcal{D}_2;$$

$$\int_{S_T} \eta[\pi^{i,\ell,1,\alpha,\gamma}, \lambda_0^{i,\ell,1,\alpha,\gamma}\varkappa^{k,i}, \mu^{i,\ell,1,\alpha,\gamma}](s,t)(w^{i,\ell,1,\alpha,\gamma} - w)\,dsdt \leq$$
$$\leq 2L[\sqrt{\gamma} + \sqrt{\mathfrak{b}^\alpha} + \sqrt{r^{i,\ell}}] \ \forall \, w \in \mathcal{D}_3;$$

$$\langle \lambda_0^{i,\ell,1,\alpha,\gamma}[-\zeta^{k,i} + \rho\frac{1+\alpha}{\alpha}|q^{k,i,\ell,\alpha,\gamma} - q^{k,i}|^{1/\alpha}h^{k,i,\ell,\alpha,\gamma}] - \sum_{j=1}^{l_k} \lambda_j^{i,\ell,1,\alpha,\gamma}e^j,$$
$$q' - q^{k,i,\ell,\alpha,\gamma}\rangle \geq -2L[\sqrt{\gamma} + \sqrt{\mathfrak{b}^\alpha} + \sqrt{r^{i,\ell}}] \ \forall \, q' \in \bar{S}_{\omega(\rho)}^{l_k}(q^{k,i});$$
$$h^{k,i,\ell,\alpha,\gamma} \in \partial|\cdot -q^{k,i}|(q^{k,i,\ell,\alpha,\gamma}), \ |h^{k,i,\ell,\alpha,\gamma}| \leq 1.$$

Here the constant $L > 0$ is defined in Lemma 3.11, and the Radon measure $\mu^{i,\ell,1,\alpha,\gamma}$ is defined by $\mu^{i,\ell,1,\alpha,\gamma} \equiv \sum_{j=1}^{l_k} \lambda_j^{i,\ell,1,\alpha,\gamma}\delta_{\tau^{k,j}}$. Going to the limit in obtained family of relations, first, when $\gamma \to 0$, and, then, as $\alpha \to \infty$, we find that there exist vectors $\lambda^{i,\ell,1} \equiv (\lambda_0^{i,\ell,1}, \lambda_1^{i,\ell,1}, \ldots, \lambda_{l_k}^{i,\ell,1}) \in R^{l_k+1}$ and

$h^{k,i,\mathfrak{k}} \in \partial |\cdot - q^{k,i}|(q^{k,i,\mathfrak{k}})$, $|h^{k,i,\mathfrak{k}}| \leq 1$, and a number $\xi^{i,\mathfrak{k}} \in [0,1]$ such that

$$\lambda_j^{i,\mathfrak{k},1} \geq 0, \quad j = \overline{0,l_k}, \quad \lambda_0^{i,\mathfrak{k},1} + \sum_{j=1}^{l_k} \lambda_j^{i,\mathfrak{k},1} = 1;$$

$$\lambda_j^{i,\mathfrak{k},1}[I_j^k(\pi^{i,\mathfrak{k},1}) - q^{k,i,\mathfrak{k}} - \varepsilon^{i,\mathfrak{k}}] = 0, \quad j = \overline{1,l_k};$$

$$\int_{Q_T} [\max_{u \in U} H(x,t,z[\pi^{i,\mathfrak{k},1}](x,t),u,\eta[\pi^{i,\mathfrak{k},1},\lambda_0^{i,\mathfrak{k},1}\varkappa^{k,i},\mu^{i,\mathfrak{k},1}](x,t)) -$$

$$- H(x,t,z[\pi^{i,\mathfrak{k},1}](x,t),u^{i,\mathfrak{k},1}(x,t),\eta[\pi^{i,\mathfrak{k},1},\lambda_0^{i,\mathfrak{k},1}\varkappa^{k,i},\mu^{i,\mathfrak{k},1}](x,t))]dxdt \leq$$

$$\leq 2L\sqrt{r^{i,\mathfrak{k}}} meas Q_T;$$

$$\int_\Omega \eta[\pi^{i,\mathfrak{k},1},\lambda_0^{i,\mathfrak{k},1}\varkappa^{k,i},\mu^{i,\mathfrak{k},1}](x,0)(v^{i,\mathfrak{k},1}(x) - v(x))\, dx +$$

$$+ \int_X \mu^{i,\mathfrak{k},1}(d\tau) \int_\Omega \nabla_v \Phi(x,\tau,z[\pi^{i,\mathfrak{k},1}](x,\tau),v^{i,\mathfrak{k},1})(v^{i,\mathfrak{k},1}(x) - v(x))\, dx \leq$$

$$\leq 2L\sqrt{r^{i,\mathfrak{k}}} \quad \forall v \in \mathcal{D}_2;$$

$$\int_{S_T} \eta[\pi^{i,\mathfrak{k},1},\lambda_0^{i,\mathfrak{k},1}\varkappa^{k,i},\mu^{i,\mathfrak{k},1}](s,t)(w^{i,\mathfrak{k},1}(s,t) - w(s,t))\, dsdt \leq 2L\sqrt{r^{i,\mathfrak{k}}}$$

$$\forall w \in \mathcal{D}_3;$$

$$\langle \lambda_0^{i,\mathfrak{k},1}[-\zeta^{k,i} + \rho\xi^{i,\mathfrak{k}}h^{k,i,\mathfrak{k}}] - \sum_{j=1}^{l_k} \lambda_j^{i,\mathfrak{k},1}e^j, q' - q^{k,i,\mathfrak{k}} \rangle \geq -2L\sqrt{r^{i,\mathfrak{k}}}$$

$$\forall q' \in \bar{S}_{\omega(\rho)}^{l_k}(q^{k,i}).$$

Here a Radon measure $\mu^{i,\mathfrak{k},1}$ is defined by $\mu^{i,\mathfrak{k},1} \equiv \sum_{j=1}^{l_k} \lambda_j^{i,\mathfrak{k},1}\delta_{\tau^{k,j}}$.

On the strength of estimates (5.13), we conclude that there exist a vector $\lambda^{i,\mathfrak{k}} \equiv \lambda^{i,\mathfrak{k},1}$, a vector $h^{k,i} \in \partial |\cdot - q^{k,i}|(q^{k,i})$, $|h^{k,i}| \leq 1$, and a sequence $\gamma^{i,\mathfrak{k}} \geq 0$, $\gamma^{i,\mathfrak{k}} \to 0$, $\mathfrak{k} \to \infty$, such that

$$\lambda_j^{i,\mathfrak{k}} \geq 0, \quad j = \overline{0,l_k}, \quad \sum_{j=0}^{l_k} \lambda_j^{i,\mathfrak{k}} = 1; \tag{5.16}$$

$$\lambda_j^{i,\mathfrak{k}} = 0 \text{ if } |I_j^k(\pi^{i,\mathfrak{k}}) - q_j^{k,i}| \geq \gamma^{i,\mathfrak{k}}, j = \overline{1, l_k};$$

$$\int_{Q_T} [\max_{u \in U} H(x, t, z[\pi^{i,\mathfrak{k}}](x, t), u, \eta[\pi^{i,\mathfrak{k}}, \lambda_0^{i,\mathfrak{k}} \varkappa^{k,i}, \mu^{i,\mathfrak{k}}](x, t)) -$$

$$- H(x, t, z[\pi^{i,\mathfrak{k}}](x, t), u^{i,\mathfrak{k}}(x, t), \eta[\pi^{i,\mathfrak{k}}, \lambda_0^{i,\mathfrak{k}} \varkappa^{k,i}, \mu^{i,\mathfrak{k}}](x, t))] dx dt \leq \gamma^{i,\mathfrak{k}};$$

$$\int_{\Omega} \eta[\pi^{i,\mathfrak{k}}, \lambda_0^{i,\mathfrak{k}} \varkappa^{k,i}, \mu^{i,\mathfrak{k}}](x, 0)(v^{i,\mathfrak{k}}(x) - v(x)) \, dx +$$

$$+ \int_X \mu^{i,\mathfrak{k}}(d\tau) \int_{\Omega} \nabla_v \Phi(x, \tau, z[\pi^{i,\mathfrak{k}}](x, \tau), v^{i,\mathfrak{k}})(v^{i,\mathfrak{k}}(x) - v(x)) \, dx \leq \gamma^{i,\mathfrak{k}}$$

$$\forall \, v \in \mathcal{D}_2;$$

$$\int_{S_T} \eta[\pi^{i,\mathfrak{k}}, \lambda_0^{i,\mathfrak{k}} \varkappa^{k,i}, \mu^{i,\mathfrak{k}}](s, t)(w^{i,\mathfrak{k}}(s, t) - w(s, t)) \, ds dt \leq \gamma^{i,\mathfrak{k}} \ \forall \, w \in \mathcal{D}_3;$$

$$\langle \lambda_0^{i,\mathfrak{k}}[\zeta^{k,i} - \rho \xi^{i,\mathfrak{k}} h^{k,i}] + \sum_{j=1}^{l_k} \lambda_j^{i,\mathfrak{k}} e^j, q' - q^{k,i} \rangle \leq \gamma^{i,\mathfrak{k}} \ \forall \, q' \in \bar{S}_{\omega(\rho)}^{l_k}(q^{k,i}).$$

Here $\mu^{i,\mathfrak{k},1} \equiv \mu^{i,\mathfrak{k}}$.

Thus, we have proved the following lemma.

Lemma 5.5 *Let* $\pi^{i,\mathfrak{k}}$, $\mathfrak{k} = 1, 2, \ldots$, *be an arbitrary m.a.s. for problem* $(P_{q^{k,i}})$ *in the sense of (1.2). Then, for any* $\rho > 0$ *there exist (depending on* ρ*) a number sequence* $\gamma^{i,\mathfrak{k}} \geq 0$, $\gamma^{i,\mathfrak{k}} \to 0$, $\mathfrak{k} \to \infty$, $\pi^{i,\mathfrak{k}} \in \mathcal{D}_{q^{k,i}}^{k,\gamma^{i,\mathfrak{k}}} \equiv \{\pi \in \mathcal{D} : I_j^k(\pi) - q_j^{k,i} \leq \gamma^{i,\mathfrak{k}}, \ j = \overline{1, l_k}\}$, *a sequence of vectors* $\lambda^{i,\mathfrak{k}} \equiv (\lambda_0^{i,\mathfrak{k}}, \lambda_1^{i,\mathfrak{k}}, \ldots, \lambda_{l_k}^{i,\mathfrak{k}}) \in R^{l_k+1}$, $\mathfrak{k} = 1, 2, \ldots$, *a number sequence* $\xi^{i,\mathfrak{k}} \in [0, 1]$, $\mathfrak{k} = 1, 2, \ldots$, *and a vector* $h^{k,i} \in \partial|\cdot - q^{k,i}|(q^{k,i})$, $|h^{k,i}| \leq 1$ *such that relations (5.16) hold, where* $\mu^{i,\mathfrak{k}} \equiv \sum_{j=1}^{l_k} \lambda_j^{i,\mathfrak{k}} \delta_{\tau^{k,j}}$, *and* $\omega(\rho) > 0$ *is a number.*

Selecting in (5.5) a subsequence $\mathfrak{k}_i$, $i = 1, 2, \ldots$, of the sequence $\mathfrak{k} = 1, 2, \ldots$, such that $\gamma^{i,\mathfrak{k}} / \omega(\rho^i) \to 0$, $i \to \infty$, where $\rho^i > 0$, $i = 1, 2, \ldots$, satisfy the conditions ρ^i, $\omega(\rho^i) \to 0$, $i \to \infty$, we obtain

$$\lambda_j^{i,\mathfrak{k}_i} \geq 0, \ j = \overline{0, l_k}, \ \sum_{j=0}^{l_k} \lambda_j^{i,\mathfrak{k}_i} = 1; \tag{5.17}$$

$$\int_{Q_T} [\max_{u \in U} H(x, t, z[\pi^{i,\mathfrak{k}_i}](x, t), u, \eta[\pi^{i,\mathfrak{k}_i}, \lambda_0^{i,\mathfrak{k}_i} \varkappa^{k,i}, \mu^{i,\mathfrak{k}_i}](x, t)) -$$

$$-H(x,t,z[\pi^{i,\mathfrak{k}_i}](x,t),u^{i,\mathfrak{k}_i}(x,t),\eta[\pi^{i,\mathfrak{k}_i},\lambda_0^{i,\mathfrak{k}_i}\varkappa^{k,i},\mu^{i,\mathfrak{k}_i}](x,t))]dxdt \le \gamma^{i,\mathfrak{k}_i};$$

$$\int_\Omega \eta[\pi^{i,\mathfrak{k}_i},\lambda_0^{i,\mathfrak{k}_i}\varkappa^{k,i},\mu^{i,\mathfrak{k}_i}](x,0)(v^{i,\mathfrak{k}_i}(x)-v(x))\,dx+$$

$$+\int_X \mu^{i,\mathfrak{k}_i}(d\tau)\int_\Omega \nabla_v\Phi(x,\tau,z[\pi^{i,\mathfrak{k}_i}](x,\tau),v^{i,\mathfrak{k}_i})(v^{i,\mathfrak{k}_i}-v)\,dx \le \gamma^{i,\mathfrak{k}_i}$$

$$\forall\, v \in \mathcal{D}_2;$$

$$\int_{S_T} \eta[\pi^{i,\mathfrak{k}_i},\lambda_0^{i,\mathfrak{k}_i}\varkappa^{k,i},\mu^{i,\mathfrak{k}_i}](s,t)(w^{i,\mathfrak{k}_i}(s,t)-w(s,t))\,dsdt \le \gamma^{i,\mathfrak{k}_i} \ \forall\, w \in \mathcal{D}_3;$$

$$\langle \lambda_0^{i,\mathfrak{k}_i}[\zeta^{k,i}-\rho^i\xi^{i,\mathfrak{k}_i}h^{k,i}]+\sum_{j=1}^{l_k}\lambda_j^{i,\mathfrak{k}_i}e_j,\delta q'\rangle \le \frac{\gamma^{i,\mathfrak{k}}}{\omega(\rho^i)} \ \forall\, \delta q' \in \bar{S}_1^{l_k}(0).$$

Let us show that a limit point of the sequence $\lambda_0^{i,\mathfrak{k}_i}$, $i=1,2,\dots$, cannot be equal to zero. Indeed, suppose that this is not true, and, without any loss of generality, $\lambda_0^{i,\mathfrak{k}_i}\to 0$, $\lambda_j^{i,\mathfrak{k}_i}\to \lambda_j$, $i\to\infty$. Going to the limit in the last relation in (5.17) as $i\to\infty$, we find that $\sum_{j=1}^{l_k}\lambda_j e^j = 0$ and, hence, vectors e^j, $j=\overline{1,l_k}$, are linearly dependent. This implies that zero cannot be a limit point of the sequence $\lambda_0^{i,\mathfrak{k}_i}$, $i=1,2,\dots$.

Setting $\pi^i \equiv \pi^{i,\mathfrak{k}_i}$, $\lambda^i \equiv \lambda^{i,\mathfrak{k}_i}/\lambda_0^{i,\mathfrak{k}_i}$, and $\mu^i \equiv \mu^{i,\mathfrak{k}_i}/\lambda_0^{i,\mathfrak{k}_i}$, we obtain

$$\lambda_j^i \ge 0, \ \ j=\overline{0,l_k}, \ \ \sum_{j=0}^{l_k}\lambda_j^i = \frac{1}{\lambda_0^{i,\mathfrak{k}_i}};$$

$$\int_{Q_T}[\max_{u\in U}H(x,t,z[\pi^i](x,t),u,\eta[\pi^i,\varkappa^{k,i},\mu^i](x,t))-$$

$$-H(x,t,z[\pi^i](x,t),u^i(x,t),\eta[\pi^i,\varkappa^{k,i},\mu^i](x,t))]dxdt \le \frac{\gamma^{i,\mathfrak{k}_i}}{\lambda_0^{i,\mathfrak{k}_i}};$$

$$\int_\Omega \eta[\pi^i,\varkappa^{k,i},\mu^i](x,0)(v^i(x)-v(x))\,dx+$$

$$+\int_X \mu^i(d\tau)\int_\Omega \nabla_v\Phi(x,\tau,z[\pi^i](x,\tau),v^i)(v^i-v)\,dx \le \frac{\gamma^{i,\mathfrak{k}_i}}{\lambda_0^{i,\mathfrak{k}_i}} \ \forall\, v \in \mathcal{D}_2;$$

$$\int_{S_T} \eta[\pi^i, \varkappa^{k,i}, \mu^i](s,t)(w^i(s,t) - w(s,t))\,dsdt \le \frac{\gamma^{i,\mathfrak{k}_i}}{\lambda_0^{i,\mathfrak{k}_i}} \ \forall w \in \mathcal{D}_3;$$

$$\langle \lambda_0^i [\zeta^{k,i} - \rho^i \xi^{i,\mathfrak{k}_i} h^{k,i}] + \sum_{j=1}^{l_k} \lambda_j^i e^j, \delta q' \rangle \le \frac{\gamma^{i,\mathfrak{k}_i}}{\lambda_0^{i,\mathfrak{k}_i} \omega(\rho^i)} \ \forall \delta q' \in \bar{S}_1^{l_k}(0).$$

Going to the limit in the last inequality when $i \to \infty$, we obtain (5.6). Then, setting $\gamma \equiv \frac{\gamma^{i,\mathfrak{k}_i}}{\lambda_0^{i,\mathfrak{k}_i}} + \varepsilon^{i,\mathfrak{k}_i}$, we obtain the other relations from lemma 5.4. Thus, lemma 5.4 is proved for the case of $\varkappa^k > 0$.

If $\varkappa^k = 0$, then, by virtue of the second formula in (5.1) for $\partial^\infty \beta_k(q^k)$, there exist sequences $\bar{\varepsilon}^i > 0$, $q^{k,i} \in R^{l_k}$, $\zeta^{k,i} \in R^{l_k}$, $\varkappa^{k,i} > 0$, $i = 1, 2, \ldots$, such that

$$(\zeta^{k,i}, -\varkappa^{k,i}) \in \hat{N}((q^{k,i}, \beta_k(q^{k,i})); \text{epi } \beta_k),$$
$$\bar{\varepsilon}^i(\zeta^{k,i}, -\varkappa^{k,i}) \to (\zeta^k, -\varkappa^k), \ q^{k,i} \to q^k, \ i \to \infty.$$

Applying reasoning similar to that in the case $\varkappa^k > 0$, we prove the assertions of lemma 5.5, and, then, the assertions of lemma 5.4. Thus, lemma 5.4 is completely proved.

Definition 5.1 *A sequence of triples $\pi^i \in \mathcal{D}$, $i = 1, 2, \ldots$, is said to be stationary in problem $(P_{q^k}^k)$ if there exist a bounded sequence of vectors $\lambda^i \equiv (\lambda_0^i, \ldots, \lambda_{l_k}^i) \in R^{l_k+1}$, $i = 1, 2, \ldots$, and a sequence of nonnegative numbers γ^i, $\gamma^i \to 0$, such that $\pi^i \in \mathcal{D}_{q^k}^{k,\gamma^i}$; relations (5.2)–(5.5) holds, where*

$$\mu^i \equiv \sum_{j=1}^{l_k} \lambda_j^i \delta_{\tau^{k,j}}, \text{ and all limit points of the sequence } \lambda^i, \ i = 1, 2, \ldots, \text{ are}$$

not equal to zero.

Definition 5.2 *A sequence $\pi^i \in \mathcal{D}$, $i = 1, 2, \ldots$, stationary in problem $(P_{q^k}^k)$ is called normal, if all limits points of any corresponding sequence λ_0^i, $i = 1, 2, \ldots$, are not equal to zero. The problem $(P_{q^k}^k)$ is said to be normal, if all stationary sequences of this problem are normal.*

A sequence $\pi^i \in \mathcal{D}$, $i = 1, 2, \ldots$, stationary in problem $(P_{q^k}^k)$ is called regular, if all limits points of some corresponding sequence λ_0^i, $i = 1, 2, \ldots$, are not equal to zero. The problem $(P_{q^k}^k)$ is said to be regular, if there exists a regular stationary sequence of this problem.

Let us introduce the following sets of multipliers: $L_{q^k}^{k,\nu} \equiv \{-\sum_{j=1}^{l_k} \lambda_j e^j \in$

$R^{l_k} : \lambda = (\lambda_0, \lambda_1, \ldots, \lambda_{l_k}) \in R^{l_k+1}, \ \lambda \neq 0, \ \lambda_0 = \nu$, there exists a sequence stationary in problem $(P_{q^k}^k)$ for which the corresponding sequence of vectors $\lambda^i, i = 1, 2, \ldots$, which is mentioned in the definition of a stationary sequence, has vector λ as its limit point}, $\nu = 0, 1$; $M_{q^k}^{k,0} \equiv L_{q^k}^{k,0} \cup \{0\}$, $M_{q^k}^{k,1} \equiv L_{q^k}^{k,1}$.

The following result immediately follows from Lemma 5.4 and the definition of generalized subdifferentials in the sense of the work [32].

Theorem 5.1 Let $\beta_k(q^k) < +\infty$. Then, $\partial \beta_k(q^k) = \partial \beta_k(q^k) \cap M_{q^k}^{k,1}$ and $\partial^\infty \beta_k(q^k) = \partial^\infty \beta_k(q^k) \cap M_{q^k}^{k,0}$, where $\partial \beta_k$ and $\partial^\infty \beta_k$ are the ordinary and singular generalized subdifferentials, respectively, in the sense of [30], [32].

¿From this theorem and [32, Corrolary 8.5] it follows that the following important result holds.

Corrolary 5.1 If, in some neighborhood O_{q^k} of a point q^k, all problems $(P_{y^k}^k)$, $y^k \in O_{q^k}$ are normal, i.e., $M_{y^k}^{k,0} = \{0\}$, $y^k \in O_{q^k}$, and the sets $M_{y^k}^{k,1}$ are uniformly bounded by a constant K in some norm $\| \cdot \|$ (for example, the Euclidean norm $| \cdot |$), then the value function β_k is Lipschitz on O_{q^k} in the norm conjugate to $\| \cdot \|$ with the same constant K.

Besides, from a lower semicontinuity of approximate problem's value function, a monotonicity of this value function with respect to each of l_k arguments, results of [39], and Lemma 5.4, we obtain the following regularity condition of problem $(P_{q^k}^k)$.

Theorem 5.2 Suppose $\hat{\partial} \beta_k(q^k) \neq \emptyset$; then the problem $(P_{q^k}^k)$ is regular. Moreover, the condition $\hat{\partial} \beta_k(q^k) \neq \emptyset$ is fulfilled for a.e. $q^k \in \mathrm{dom}\, \beta_k$.

6. Proof of Regularity and Normality Conditions

The proof of Theorem 2.2. Firstly, for brevity, let us introduce the following Weierstrass E–functions: $E_G \colon \Omega \times R \times R \to R$, $E_\Phi \colon \Omega \times [0, T] \times R \times R \times V \times V \to R$, $E_a \colon Q_T \times R \times R \times U \times U \to R$. Namely, by definition, put $E_a(x, t, z_1, z_2, u_1, u_2) \equiv a(x, t, z_1, u_1) - a(x, t, z_2, u_1) - \nabla_z a(x, t, z_2, u_2)(z_1 - z_2)$, $E_\Phi(x, z_1, z_2, v_1, v_2) \equiv \Phi(x, z_1, v_1) - \Phi(x, z_2, v_2) - \nabla_z \Phi(x, z_2, v_2)(z_1 - z_2) - \nabla_v \Phi(x, z_2, v_2)(v_1 - v_2)$, $E_G(x, z_1, z_2) \equiv G(x, z_1) - G(x, z_2) - \nabla_z G(x, z_2)(z_1 - z_2)$. These functions are analogues of functions of calculus of variation with the same names. The E–functions were introduced in optimal control theory by V.I. Plotnikov [40] (see also [41],[42]).

Using these functions and Lemma 3.8, it can be shown that increments of functionals $I_0(\cdot)$ and $I_1(\cdot)(\tau)$, $\tau \in [0,T]$, can be represented in the form (see, e.g., [42])

$$\Delta I_0 \equiv I_0(\pi^1) - I_0(\pi^2) \equiv \tag{6.1}$$

$$\equiv \left\{ \int_\Omega E_G(x, z_1(x,T), z_2(x,T))dx - \int_{Q_T} E_a(x,t,z_1,z_2,u^1,u^2)\mathfrak{p}_0[\pi^2](x,t)dxdt \right\} +$$

$$+ \left\{ \int_\Omega \mathfrak{p}_0[\pi^2](x,0)(v^1(x)-v^2(x))dx + \int_{S_T} \mathfrak{p}_0[\pi^2](s,t)(w^1(s,t)-w^2(s,t))dsdt - \right.$$

$$\left. - \int_{Q_T} [H(x,t,z_2,u^1(x,t),\mathfrak{p}_0[\pi^2](x,t)) - H(x,t,z_2,u^2(x,t),\mathfrak{p}_0[\pi^2](x,t))]dxdt \right\} \equiv$$

$$\equiv \{\mathcal{E}_{I_0}(\pi^1,\pi^2)\} + \{\mathcal{H}_{I_0}(\pi^1,\pi^2)\};$$
$$\Delta I_1(\tau) \equiv I_1(\pi^1)(\tau) - I_1(\pi^2)(\tau) \equiv$$

$$\equiv \left\{ \int_\Omega E_\Phi(x,\tau,z_1(x,\tau),z_2(x,\tau),v^1,v^2)dx - \right.$$

$$\left. - \int_{Q_T} E_a(x,t,z_1,z_2,u^1,u^2)\mathfrak{p}_1[\pi^2](x,t,\tau)dxdt \right\} + \left\{ \int_\Omega [\mathfrak{p}_1[\pi^2](x,0,\tau) + \right.$$

$$+\nabla_v \Phi(x,z_2(x,\tau),v^2(x))](v^1(x)-v^2(x))dx + \int_{S_T} \mathfrak{p}_1[\pi^2](s,t,\tau)(w^1-w^2)dsdt -$$

$$\left. - \int_{Q_T} [H(x,t,z_2,u^1,\mathfrak{p}_1[\pi^2](x,t,\tau)) - H(x,t,z_2,u^2,\mathfrak{p}_1[\pi^2](x,t,\tau))]dxdt \right\} \equiv$$

$$\equiv \{\mathcal{E}_{I_1}(\pi^1,\pi^2)(\tau)\} + \{\mathcal{H}_{I_1}(\pi^1,\pi^2)(\tau)\}.$$

Here $\pi^i \equiv (u^i,v^i,w^i) \in \mathcal{D}$, $i=1,2$, are arbitrary, and $z_i \equiv z[\pi^i]$, $i=1,2$.

Suppose a sequence of triples $\pi^i \equiv (u^i,v^i,w^i) \in \mathcal{D}$, $i=1,2,\ldots$, is a regular stationary sequence in problem (P_q); then this sequence satisfies the maximum principle relations for some bounded sequence of pairs $(\lambda^i,\mu^i) \in R \times M(X)$, $i=1,2,\ldots$, and all limit points of the sequence λ^i, $i=1,2,\ldots$, are not equal to zero. Let $\pi \equiv (u,v,w) \in \mathcal{D}$ be arbitrary. Consider the expression $\lambda^i(I_0(\pi)-I_0(\pi^i)) + \int_X (I_1(\pi)(\tau)-I_1(\pi^i)(\tau))\mu^i(d\tau)$. According to equalities (6.1),

$$\lambda^i(I_0(\pi) - I_0(\pi^i)) + \int_X (I_1(\pi)(\tau) - I_1(\pi^i)(\tau))\mu^i(d\tau) = \Big\{\lambda^i \mathcal{E}_{I_0}(\pi, \pi^i) +$$

$$+ \int_X \mathcal{E}_{I_1}(\pi, \pi^i)(\tau)\mu^i(d\tau)\Big\} + \Big\{\lambda^i \mathcal{H}_{I_0}(\pi, \pi^i) + \int_X \mathcal{H}_{I_1}(\pi, \pi^i)(\tau)\mu^i(d\tau)\Big\} \equiv$$

$$\equiv \{\mathcal{E}(\pi, \pi^i)\} + \{\mathcal{H}(\pi, \pi^i)\},$$

where $\mathcal{E}(\pi, \pi^i)$ is a summary generalized Weierstrass–Plotnikov E–function.

In view of a linear convexity of problem (P_q), we have $\mathcal{E}(\pi, \pi^i) \geq 0, \forall \pi \in \mathcal{D}$. It follows that

$$\lambda^i(I_0(\pi) - I_0(\pi^i)) + \int_X (I_1(\pi)(\tau) - I_1(\pi^i)(\tau))\mu^i(d\tau) =$$

$$= \mathcal{E}(\pi, \pi^i) + \mathcal{H}(\pi, \pi^i) \geq \mathcal{H}(\pi, \pi^i).$$

Since $\pi^i \in \mathcal{D}$, $i = 1, 2, \ldots$, is regular stationary sequence, the sequence of numbers λ^i, $i = 1, 2, \ldots$, is uniformly bounded from below by some positive number. From this fact, the last inequality, and a stationary of the sequence $\pi^i \equiv (u^i, v^i, w^i) \in \mathcal{D}$, $i = 1, 2, \ldots$, we conclude that there exist a sequence of nonnegative numbers $\varepsilon^i \to 0$, $i \to \infty$, such that

$$-\varepsilon^i \leq \lambda^i[I_0(\pi) - I_0(\pi^i)] + \int_X [I_1(\pi)(\tau) - I_1(\pi^i)(\tau)]\mu^i(d\tau) \qquad (6.2)$$

$$\forall \pi \in \mathcal{D}.$$

Moreover, since the sequence $\pi^i \in \mathcal{D}$, $i = 1, 2, \ldots$, is a stationary sequence, there exist a sequence of nonnegative numbers $\gamma^i \to 0$, $i \to \infty$, such that $\pi^i \in \mathcal{D}_q^{\gamma^i}$, and the measure μ^i is concentrated on the set $\{\tau \in X : |I_1(\pi^i)(\tau) - q(\tau)| \leq \gamma^i\}$, $i = 1, 2, \ldots$.

Under the condition $\pi \in \mathcal{D}_q^{\gamma^i}$, let us estimate the right–hand side part of (6.2):

$$-\varepsilon^i \leq \lambda^i(I_0(\pi) - I_0(\pi^i)) + \int_X (I_1(\pi)(\tau) - I_1(\pi^i)(\tau))\mu^i(d\tau) =$$

$$= \lambda^i[I_0(\pi) - I_0(\pi^i)] + \int_X [I_1(\pi)(\tau) - q(\tau)]\mu^i(d\tau) +$$

$$+ \int_X [q(\tau) - I_1(\pi^i)(\tau)]\mu^i(d\tau) \le \lambda^i[I_0(\pi) - I_0(\pi^i)] + \gamma^i\|\mu^i\| +$$

$$+ \int_X |q(\tau) - I_1(\pi^i)(\tau)|\mu^i(d\tau) \le \lambda^i[I_0(\pi) - I_0(\pi^i)] + \gamma^i\|\mu^i\| + \gamma^i\|\mu^i\| =$$

$$= \lambda^i[I_0(\pi) - I_0(\pi^i)] + 2\gamma^i\|\mu^i\|.$$

Thus, $-(\varepsilon^i + 2\gamma^i\|\mu^i\|) \le \lambda^i(I_0(\pi) - I_0(\pi^i))$. Therefore,

$$I_0(\pi^i) \le I_0(\pi) + \frac{\varepsilon^i + 2\gamma^i\|\mu^i\|}{\lambda^i} \quad \forall \pi \in \mathcal{D}_q^{\gamma^i}.$$

Hence,

$$I_0(\pi^i) \le \beta_{\gamma^i}(q) + \frac{\varepsilon^i + 2\gamma^i\|\mu^i\|}{\lambda^i} \le \beta(q) + \frac{\varepsilon^i + 2\gamma^i\|\mu^i\|}{\lambda^i}.$$

From this inequality, inclusions $\pi^i \in \mathcal{D}_q^{\gamma^i}$, $i = 1, 2, \ldots$, and m.a.s. definition, it follows that $\pi^i \in \mathcal{D}$, $i = 1, 2, \ldots$, is a m.a.s. in problem (P_q). This completes the proof of theorem 2.2.

The proof of Theorem 2.3. The proof is in two steps.

Step 1. Let us prove the first assertion of the theorem. Assume the converse. Let $\pi^i \equiv (u^i, v^i, w^i) \in \mathcal{D}$, $i = 1, 2, \ldots$, be a stationary sequence in problem (P_q) such that the corresponding sequence of pairs (λ^i, μ^i), $i = 1, 2, \ldots$, has a limit point (λ, μ) with $\lambda = 0$. Suppose i_k, $k = 1, 2, \ldots$, is a subsequence of the sequence $i = 1, 2, \ldots$, such that $\lambda^{i_k} \to \lambda$, $k \to \infty$, $\mu^{i_k} \to \mu$, $k \to \infty$, $*$–weakly; then[‡]

$$\lambda^{i_k}[I_0(\pi_0) - I_0(\pi^{i_k})] + \int_X [I_1(\pi_0)(\tau) - I_1(\pi^{i_k})(\tau)]\mu^{i_k}(d\tau) =$$

$$\lambda^{i_k}[\mathcal{E}_{I_0}(\pi_0, \pi^{i_k}) + \mathcal{H}_{I_0}(\pi_0, \pi^{i_k})] + \int_X [\mathcal{E}_{I_1}(\pi_0, \pi^{i_k})(\tau) +$$

$$+ \mathcal{H}_{I_1}(\pi_0, \pi^{i_k})(\tau)]\mu^{i_k}(d\tau) \ge \lambda^{i_k}\mathcal{E}_{I_0}(\pi_0, \pi^{i_k}) +$$

$$+ \lambda^{i_k}\mathcal{H}_{I_0}(\pi_0, \pi^{i_k}) + \int_X \mathcal{H}_{I_1}(\pi_0, \pi^{i_k})(\tau)\mu^{i_k}(d\tau).$$

[‡]Notation $\mathcal{E}_{I_0}, \mathcal{H}_{I_0}, \mathcal{E}_{I_1}, \mathcal{H}_{I_1}$ are introduced in the proof of Theorem 2.2 above.

Hence, in view of a stationarity of the sequence $\pi^i \in \mathcal{D}$, $i = 1, 2, \ldots$, there exists a sequence of nonnegative numbers $\gamma^{ik} \to 0$, $k \to \infty$, such that

$$-\gamma^{ik} \le \lambda^{ik}\mathcal{H}_{I_0}(\pi_0, \pi^{ik}) + \int_X \mathcal{H}_{I_1}(\pi_0, \pi^{ik})(\tau)\mu^{ik}(d\tau) \le -\lambda^{ik}\mathcal{E}_{I_0}(\pi_0, \pi^{ik}) +$$

$$+\lambda^{ik}[I_0(\pi_0) - I_0(\pi^{ik})] + \int_X [I_1(\pi_0)(\tau) - q(\tau)]\mu^{ik}(d\tau) +$$

$$+\int_X [q(\tau) - I_1(\pi^{ik})(\tau)]\mu^{ik}(d\tau) \le [-\lambda^{ik}\mathcal{E}_{I_0}(\pi_0, \pi^{ik}) + \lambda^{ik}[I_0(\pi_0) -$$

$$-I_0(\pi^{ik})]] + \int_X [I_1(\pi_0)(\tau) - q(\tau)]\mu^{ik}(d\tau) + \gamma^{ik}\|\mu^{ik}\| \le -\varepsilon < 0,$$

for all $k \ge k_0$, where k_0 is some natural number, $\varepsilon > 0$ is some number. But $\gamma^{ik} \to 0$, $k \to \infty$. This contradiction proves that problem (P_q) is normal.

Step 2. Let us prove the second assertion of the theorem. Suppose this is not true. Let $\bar\pi^i \equiv (\bar u^i, \bar v^i, \bar w^i) \in \mathcal{D}_q^{\bar\gamma^i}$, $i = 1, 2, \ldots$, be a stationary sequence in problem (P_q) such that a corresponding sequence of pairs (λ^i, μ^i), $i = 1, 2, \ldots$, has a limit point (λ, μ) with $\lambda = 0$. Without loss of generality it can be assumed that $\lambda^i \to \lambda$, $i \to \infty$, $\mu^i \to \mu$, $i \to \infty$, $*$–weakly. Suppose a sequence $\pi_*^i \equiv (u_*^i, v_*^i, w_*^i) \in \mathcal{D}_q^{\bar\gamma^i}$, $i = 1, 2, \ldots$, is arbitrary, where $\bar\gamma^i \ge 0$, $\bar\gamma^i \to 0$, $i \to \infty$; then $\lambda^i[I_0(\pi_*^i) - I_0(\bar\pi^i)] + \int_X [I_1(\pi_*^i)(\tau) - I_1(\bar\pi^i)(\tau)]\mu^i(d\tau) = \mathcal{H}(\pi_*^i, \bar\pi^i)$. Therefore,

$$-\tilde\gamma^i \le \mathcal{H}(\pi_*^i, \bar\pi^i) = \lambda^i[I_0(\pi_*^i) - I_0(\bar\pi^i)] + \int_X [I_1(\pi_*^i)(\tau) -$$

$$-I_1(\bar\pi^i)(\tau)]\mu^i(d\tau) = \lambda^i[I_0(\pi_*^i) - I_0(\bar\pi^i)] + \int_X [I_1(\pi_*^i)(\tau) - q(\tau)]\mu^i(d\tau) +$$

$$+\int_X [q(\tau) - I_1(\bar\pi^i)(\tau)]\mu^i(d\tau) \le \{\lambda^i[I_0(\pi_*^i) - I_0(\bar\pi^i)] + \bar C[\tilde\gamma^i + \gamma^i]] \equiv \bar\beta^i.$$

Hence, $\mathcal{H}(\pi_*^i, \bar\pi^i) \equiv \bar\beta_1^i \to 0$, $i \to \infty$. Besides, by virtue of assumptions of the theorem, functions $\mathfrak{p}_0$, $\mathfrak{p}_1$ do not depend on $\pi \in \mathcal{D}$. Using this fact, the convergence $\lambda^i \to 0$, $i \to \infty$, and a stationarity of the sequence $\bar\pi^i \equiv (\bar u^i, \bar v^i, \bar w^i) \in \mathcal{D}_q^{\bar\gamma^i}$, $i = 1, 2, \ldots$, we obtain that there exists a sequence $\bar\beta_2^i \ge 0$, $\bar\beta_2^i \to 0$, $i \to \infty$, such that

$$\mathcal{H}(\pi_*^i, \bar\pi^i) \ge -\bar\beta_2^i, \quad -\bar\beta_2^i \le \int_X [I_1(\pi_*^i)(\tau) - q(\tau)]\mu^i(d\tau) \le \bar C\tilde\gamma^i.$$

From inclusions $\pi_*^i \in \mathcal{D}_q^{\bar{\gamma}^i}$, $i = 1, 2, \ldots$, it follows that problem (P_q) has not nonstationary sequence. This contradiction proves that problem (P_q) is normal.

The proof of Theorem 2.4. Let us introduce a functional $\mathcal{J}^\gamma \colon \mathcal{D} \to R$ by
$\mathcal{J}^\gamma(\pi) \equiv \max\{I_0(\pi) - \beta_0(q) + \gamma - \varepsilon^i; \ \max\limits_{\tau \in X}[I_1(\pi)(\tau) - q(\tau)] - \varepsilon^i\}$, where $\gamma > 0$ is a parameter. Let us show that $\mathcal{J}^\gamma(\pi) \geq 0 \ \forall \pi \in \mathcal{D}$. Indeed, suppose there exists $\pi^* \in \mathcal{D}$ such that $\mathcal{J}^\gamma(\pi^*) < 0$. Then there exists a number $\vartheta > 0$ such that $\mathcal{J}^\gamma(\pi^*) \leq -\vartheta$. Hence, $I_0(\pi^*) \leq \beta_0(q) + \varepsilon^i - \gamma - \vartheta$, and $\max\limits_{\tau \in X}[I_1(\pi^*)(\tau) - q(\tau)] \leq -\vartheta + \varepsilon^i$. Because $\varepsilon^i \to 0$, $i \to \infty$, there exists a number i_0 such that $\varepsilon^i \leq \frac{\vartheta}{2}$ for all $i \geq i_0$. Then for any such i $I_0(\pi^*) \leq \beta_0(q) - \gamma - \frac{\vartheta}{2} < \beta_0(q)$ and $\max\limits_{\tau \in X}[I_1(\pi^*)(\tau) - q(\tau)] \leq -\frac{\vartheta}{2} < 0$. Thus, $\pi^* \in \mathcal{D}_q^0$ and $I_0(\pi^*) < \beta_0(q) \equiv \inf\limits_{\pi \in \mathcal{D}_q^0} I_0(\pi)$. It is impossible. Thus, $\mathcal{J}^\gamma(\pi) \geq 0 \ \forall \pi \in \mathcal{D}$.

Since $\mathcal{J}^\gamma(\pi) \geq 0 \ \forall \pi \in \mathcal{D}$, the following relation holds: $\mathcal{J}^\gamma(\pi^i) \leq \inf\limits_{\pi \in \mathcal{D}} \mathcal{J}^\gamma(\pi) + \gamma$. From this inequality and the Ekeland variational principle [33], it follows that there exists a triple $\pi^{i,\gamma} \in \mathcal{D}$ that is a solution to the problem

$$\mathcal{J}^\gamma(\pi) + \sqrt{\gamma} d(\pi, \pi^{i,\gamma}) \to \min, \quad \pi \in \mathcal{D},$$

and satisfies the inequalities

$$d(\pi^i, \pi^{i,\gamma}) \leq \sqrt{\gamma}, \quad \mathcal{J}^\gamma(\pi^{i,\gamma}) \leq \mathcal{J}^\gamma(\pi^i).$$

Consider optimization problems families

$$\mathcal{J}_k^\gamma(\pi) + \sqrt{\gamma} d(\pi, \pi^{i,\gamma}) \to \min, \quad \pi \in \mathcal{D},$$

where $\mathcal{J}_k^\gamma(\pi) \equiv \max\{I_0(\pi) - \beta_0(q) + \gamma - \varepsilon^i; \ \max\limits_{\tau \in \hat{X}^k}[I_1(\pi)(\tau) - q(\tau)] - \varepsilon^i\}$, $k = 1, 2, \ldots$. Here $\hat{X}^k \subseteq X$ is a finite $1/k$–net of a compact set X. This net is defined in the section 5.. In view of precompactness of the operator I_1 image in $C(X)$,

$$\lim\limits_{k \to \infty} \inf\limits_{\pi \in \mathcal{D}}[\mathcal{J}_k^\gamma(\pi) + \sqrt{\gamma} d(\pi, \pi^{i,\gamma})] = \inf\limits_{\pi \in \mathcal{D}}[\mathcal{J}^\gamma(\pi) + \sqrt{\gamma} d(\pi, \pi^{i,\gamma})] = \mathcal{J}^\gamma(\pi^{i,\gamma}).$$

Therefore, there exists a sequence $\varkappa^{i,\gamma,k} \geq 0$, $k = 1, 2, \ldots$, $\varkappa^{i,\gamma,k} \to 0$, $k \to \infty$, such that

$$\mathcal{J}_k^\gamma(\pi^{i,\gamma}) = \mathcal{J}_k^\gamma(\pi^{i,\gamma}) + \sqrt{\gamma} d(\pi^{i,\gamma}, \pi^{i,\gamma}) \leq$$
$$\leq \inf\limits_{\pi \in \mathcal{D}}[\mathcal{J}_k^\gamma(\pi) + \sqrt{\gamma} d(\pi, \pi^{i,\gamma})] + \varkappa^{i,\gamma,k}.$$

Applying again the Ekeland variational principle [33], we conclude that there exists a sequence $\pi^{i,\gamma,k} \in \mathcal{D}$, $k = 1, 2, \ldots$, such that for any $k = 1, 2, \ldots$, triple $\pi^{i,\gamma,k}$ is a solution to the problem

$$\mathcal{J}_k^\gamma(\pi) + \sqrt{\gamma}d(\pi, \pi^{i,\gamma}) + \sqrt{\varkappa^{i,\gamma,k}}d(\pi, \pi^{i,\gamma,k}) \to \min, \quad \pi \in \mathcal{D},$$

and satisfies the inequalities

$$d(\pi^{i,\gamma}, \pi^{i,\gamma,k}) \leq \sqrt{\varkappa^{i,\gamma,k}}, \quad \mathcal{J}_k^\gamma(\pi^{i,\gamma,k}) + \sqrt{\gamma}d(\pi^{i,\gamma,k}, \pi^{i,\gamma}) \leq \mathcal{J}^\gamma(\pi^{i,\gamma}).$$

Applying reasoning similar to that in the proof of the relations (4.10)–(4.14), we obtain that there exist a nonnegative numbers sequence $\lambda_0^{i,\gamma,k}$, $k = 1, 2, \ldots$, and a sequence of nonnegative Radon measures $\mu^{i,\gamma,k} \in M(X)$, $k = 1, 2, \ldots$, such that

$$\lambda_0^{i,\gamma,k} + \|\mu^{i,\gamma,k}\| = 1, \quad \lambda_0^{i,\gamma,k}[\mathcal{J}_k^\gamma(\pi^{i,\gamma,k}) - [I_0(\pi^{i,\gamma,k}) - \beta_0(q) + \gamma - \varepsilon^i]] = 0;$$

$$\|\mu^{i,\gamma,k}\|J_k(\pi^{i,\gamma,k}) - \int_X [I_1(\pi^{i,\gamma,k})(\tau) - q(\tau) - \varepsilon^i]\mu^{i,\gamma,k}(d\tau) = 0;$$

$$H(x, t, z[\pi^{i,\gamma,k}](x,t), u, \eta[\pi^{i,\gamma,k}, \lambda_0^{i,\gamma,k}, \mu^{i,\gamma,k}](x,t)) -$$

$$- H(x, t, z[\pi^{i,\gamma,k}](x,t), u^{i,\gamma,k}(x,t), \eta[\pi^{i,\gamma,k}, \lambda_0^{i,\gamma,k}, \mu^{i,\gamma,k}](x,t)) \leq$$

$$\leq 2L[\sqrt{\gamma} + \sqrt{\varkappa^{i,\gamma,k}}] \; \forall \, u \in U \text{ for a.e. } (x, t) \in Q_T;$$

$$\int_\Omega \eta[\pi^{i,\gamma,k}, \lambda_0^{i,\gamma,k}, \mu^{i,\gamma,k}](x,0)(v^{i,\gamma,k}(x) - \tilde{v}(x)) \, dx +$$

$$+ \int_X \mu^{i,\gamma,k}(d\tau) \int_\Omega \nabla_v \Phi(x, \tau, z[\pi^{i,\gamma,k}](x,\tau), v^{i,\gamma,k}(x))(v^{i,\gamma,k} - \tilde{v}) \, dx \leq$$

$$\leq 2L[\sqrt{\gamma} + \sqrt{\varkappa^{i,\gamma,k}}] \; \forall \, \tilde{v} \in \mathcal{D}_2;$$

$$\int_{S_T} \eta[\pi^{i,\gamma,k}, \lambda_0^{i,\gamma,k}, \mu^{i,\gamma,k}](s,t)(w^{i,\gamma,k}(s,t) - \tilde{w}(s,t)) \, dsdt \leq$$

$$\leq 2L[\sqrt{\gamma} + \sqrt{\varkappa^{i,\gamma,k}}] \; \forall \, \tilde{w} \in \mathcal{D}_3.$$

Going to the limit in these relations, first when $k \to \infty$, and, then, as $\gamma \to 0$, we obtain that there exist a sequence of nonnegative numbers λ_0^i, $i = 1, 2, \ldots$; a sequence of nonnegative numbers $\mathfrak{b}^i$, $i = 1, 2, \ldots$, $\mathfrak{b}^i \to 0$, $i \to \infty$; and a sequence of nonnegative Radon measures $\mu^i \in M(X)$, $i = 1, 2, \ldots$, ($\mu^i \in$

$M(X)$ is concentrated on $\{\tau \in X : |I_1(\pi^i)(\tau) - q(\tau)| \leq \mathfrak{b}^i\}$) such that

$$\lambda_0^i + \|\mu^i\| = 1, \quad \lambda_0^i[I_0(\pi^i) - \beta_0(q) - \varepsilon^i] = 0;$$

$$\int_{Q_T} [\max_{u \in U} H(x, t, z[\pi^i](x, t), u, \eta[\pi^i, \lambda_0^i, \mu^i](x, t)) -$$

$$- H(x, t, z[\pi^i](x, t), u^i(x, t), \eta[\pi^i, \lambda_0^i, \mu^i](x, t))] dx dt \leq \mathfrak{b}^i;$$

$$\max_{v \in \mathcal{D}_2} \left\{ \int_{\Omega} \eta[\pi^i, \lambda_0^i, \mu^i](x, 0)(v^i(x) - v(x)) \, dx + \right.$$

$$\left. + \int_X \mu^i(d\tau) \int_{\Omega} \nabla_v \Phi(x, \tau, z[\pi^i](x, \tau), v^i(x))(v^i(x) - v(x)) \, dx \right\} \leq \mathfrak{b}^i;$$

$$\max_{w \in \mathcal{D}_3} \int_{S_T} \eta[\pi^i, \lambda_0^i, \mu^i](s, t)(w^i(s, t) - w(s, t)) \, ds dt \leq \mathfrak{b}^i.$$

In other words, the sequence π^i, $i = 1, 2, \ldots$, is a stationary sequence in problem (P_q). In addition, if $I_0(\pi^i) \to \bar{\beta} \in [\beta(q), \beta_0(q))$, $i \to \infty$, then, since $\varepsilon^i \to 0$, $i \to \infty$, we have $I_0(\pi^i) - \beta_0(q) - \varepsilon^i < 0$ for all enough large i. Hence, $\lambda_0^i = 0$ for all enough large i. Thus, if $I_0(\pi^i) \to \bar{\beta} \in [\beta(q), \beta_0(q))$, $i \to \infty$, then π^i, $i = 1, 2, \ldots$, is a stationary (but not normal!) sequence in problem (P_q). This completes the proof.

The proof of Theorem 2.5. Suppose that this is not true; i.e., problem (P_q) is normal and $\beta(q) < \beta_0(q)$. Let π^i, $i = 1, 2, \ldots$, be a stationary sequence in problem (P_q), and let $I_0(\pi^i) \to \beta(q)$, $i \to \infty$. ¿From the theorem 2.4 and a strict inequality $\beta(q) < \beta_0(q)$, it follows that π^i, $i = 1, 2, \ldots$, is a stationary (but not normal!) sequence in problem (P_q). This contradicts normality of problem (P_q). Theorem is proved.

The proof of Theorem 2.6. Let us show that the normality of problem (P_q) implies that its value function is Lipschitz in a neighborhood of the point $q \in C(X)$. To this end, we first prove that if problem (P_q) is normal, then there exists $\delta > 0$ such that all sets $M_{y^k}^{k,1} \subset R^{l_k}$, $k = 1, 2, \ldots$, for $|y^k - \bar{q}^k|_\infty \leq \delta$, are uniformly bounded with respect to $k = 1, 2, \ldots$ and y^k in the 1–norm $|\cdot|_1$, where the 1–norm $|x|_1$ is meant to be a number $\sum_{i=1}^{l_k} |x_i|$. Suppose that this is not true; i.e., there exist sequences of vectors $\tilde{y}^k \in R^{l_k}$, $\lambda^k \in \hat{M}_{y^k}^{k,1}$, $k = 1, 2, \ldots$, such that $|\tilde{y}^k - \bar{q}^k|_\infty \to 0$, $|\lambda^k|_1 \to \infty$, $k \to \infty$.

This implies that, for any $k = 1, 2, \ldots$, there exist a sequence of pairs;

$\pi^{k,i}$, $i = 1, 2, \ldots$; a sequence of numbers $\gamma^{k,i} \geq 0$, $\gamma^{k,i} \to 0$, $i \to \infty$; $\pi^{k,i} \in \mathcal{D}_{\tilde{y}^k}^{k,\gamma^{k,i}}$, $i = 1, 2, \ldots$; and a bounded sequence of vectors $\lambda^{k,i} \equiv (\lambda_0^{k,i}, \ldots, \lambda_{l_k}^{k,i}) \in R^{l_k+1}$, $i = 1, 2, \ldots$, such that

$$\int_{Q_T} [\max_{u' \in U} H(x, t, z[\pi^{k,i}](x,t), u', \eta[\pi^{k,i}, \lambda_0^{k,i}, \mu^{k,i}](x,t)) -$$

$$- H(x, t, z[\pi^{k,i}](x,t), u^{k,i}(x,t), \eta[\pi^{k,i}, \lambda_0^{k,i}, \mu^{k,i}](x,t))] \, dxdt \leq \gamma^{k,i},$$

$$\max_{v \in \mathcal{D}_2} \left\{ \int_\Omega \eta[\pi^{k,i}, \lambda_0^{k,i}, \mu^{k,i}](x,0)(v^{k,i}(x) - v(x)) \, dx + \right.$$

$$\left. + \int_X \mu^{k,i}(d\tau) \int_\Omega \nabla_v \Phi(x, \tau, z[\pi^{k,i}](x,\tau), v^{k,i}(x))(v^{k,i} - v) \, dx \right\} \leq \gamma^{k,i},$$

$$\max_{w \in \mathcal{D}_2} \int_{S_T} \eta[\pi^{k,i}, \lambda_0^{k,i}, \mu^{k,i}](s,t)(w^{k,i} - w) \, dsdt \leq \gamma^{k,i}, \quad i = 1, 2, \ldots,$$

where $\mu^{k,i} \equiv \sum_{j=1}^{l_k} \lambda_j^{k,i} \delta_{\tau^{k,j}}$.

Let i_k, $k = 1, 2, \ldots$, be a sequence such that $\gamma^{k,i_k} \to 0$, $\gamma^{k,i_k}/|\lambda^k|_1 \equiv \bar{\gamma}^k \to 0$, $\lambda_0^{k,i_k}/|\lambda^k|_1 \equiv \tilde{\lambda}_0^k \to 0$, $|\tilde{\lambda}^{k,i_k}/|\lambda^k|_1|_1 \to 1$, $\tilde{\lambda}^{k,i_k}/|\lambda^k|_1 \equiv \tilde{\lambda}^k$. Let also $\tilde{\mu}^k \equiv \sum_{j=1}^{l_k} \tilde{\lambda}_j^k \delta_{\tau^{k,j}}$. Then the sequence $\pi^k \equiv \pi^{k,i_k} \in \mathcal{D}_q^{\bar{\gamma}_1^k}$, $k = 1, 2, \ldots$, $\bar{\gamma}_1^k \geq 0$, $\bar{\gamma}_1^k \to 0$, $k \to \infty$, satisfies the inequalities

$$\int_{Q_T} [\max_{u' \in U} H(x, t, z[\pi^k](x,t), u', \eta[\pi^k, \tilde{\lambda}_0^k, \tilde{\mu}^k](x,t)) -$$

$$- H(x, t, z[\pi^k](x,t), u^k(x,t), \eta[\pi^k, \tilde{\lambda}_0^k, \tilde{\mu}^k](x,t))] \, dxdt \leq \bar{\gamma}_1^k,$$

$$\max_{v \in \mathcal{D}_2} \left\{ \int_\Omega \eta[\pi^k, \tilde{\lambda}_0^k, \tilde{\mu}^k](x,0))(v^k(x) - v(x)) \, dx + \right.$$

$$\left. + \int_X \tilde{\mu}^k(d\tau) \int_\Omega \nabla_v \Phi(x, \tau, z[\pi^k](x,\tau), v^k(x))(v^k(x) - v(x)) \, dx \right\} \leq \bar{\gamma}_1^k,$$

$$\max_{w \in \mathcal{D}_2} \int_{S_T} \eta[\pi^k, \tilde{\lambda}_0^k, \tilde{\mu}^k](s,t)(w^k(s,t) - w(s,t)) \, dsdt \leq \bar{\gamma}_1^k, \quad k = 1, 2, \ldots,$$

where, obviously, $\tilde{\lambda}_0^k \to 0$, $\|\tilde{\lambda}^k\| \to 1$, $k \to \infty$, and a nonnegative measure $\tilde{\mu}^k \in M(X)$ is concentrated on the set $\{\tau \in X \, : \, |I_1(\pi^k)(\tau) - q(\tau)| \le \bar{\gamma}_2^k\}$, $\bar{\gamma}_2^k \to 0$, $k \to \infty$. Here, $\bar{\gamma}_s^k$, $k = 1, 2, \ldots$, $s = 1, 2$, are some sequences of nonnegative numbers. Thus, π^k, $k = 1, 2, \ldots$, is a stationary (but not normal!) sequence in problem (P_q).

The contradiction we arrived at implies that the sets $M_{y^k}^{k,1}$ are actually uniformly bounded. Moreover, it can be shown without loss of generality that all problems $(P_{y^k}^k)$ are normal if y^k satisfies the inequality $|y^k - \bar{q}^k|_\infty \le \delta$. Hence, according to corrolary 5.1, the value functions $\beta_k(y^k)$ are Lipschitz for $|y^k - \bar{q}^k|_\infty \le \delta/2$, with the constant $\bar{K}$ being independent of $k = 1, 2, \ldots$. Then, it follows that the value function of the original problem (P_q) also satisfies the Lipschitz condition with the same constant in the $\delta/2$–neighborhood of the point $q \in C(X)$. Indeed, let $q^i \in C(X)$, $|q^i - q|_X^{(0)} \le \delta/2$, $i = 1, 2$. Then the inequalities $|\bar{q}^{i,k} - \bar{q}^k|_\infty \le \delta/2$, $i = 1, 2$, also hold, where $\bar{q}^{i,k} \equiv (q^i(\tau^{k,1}), \ldots, q^i(\tau^{k,l_k}))$, $i = 1, 2$, $k = 1, 2, \ldots$ ¿From these inequalities, it follows that $|\beta_k(\bar{q}^{1,k}) - \beta_k(\bar{q}^{2,k})| \le \bar{K}|\bar{q}^{1,k} - \bar{q}^{2,k}|_\infty$. Going to the limit in the last inequality as $k \to \infty$ and taking into account lemma 5.1, we obtain $|\beta(q^1) - \beta(q^2)| \le \bar{K}|q^1 - q^2|_X^{(0)}$. The theorem is proved.

The proof of Theorem 2.7. Let us give a main idea of the proof. Consider the family of optimization problems $(\bar{P}_\rho) \equiv (P_{q+\rho\tilde{q}})$ depending on a number parameter ρ, where $\tilde{q} \equiv 1$. Because the function β is Lipschitz in a neighborhood of the point q, then, obviously, the one variable function $\bar{\beta}(\rho) \equiv \beta(q + \rho\tilde{q})$ is Lipschitz in a neighborhood of zero. Hence, by virtue of the first formula (5.1), there exist sequences of numbers $\rho^i, \zeta^i, \varkappa^i, i = 1, 2, \ldots$, such that

$$\rho^i \to 0, \ \ \bar{\beta}(\rho^i) \to \bar{\beta}(0), (\zeta^i, -\varkappa^i) \to (\zeta, -\varkappa) \ne 0, \ \ i \to \infty, \ \ \varkappa > 0,$$
$$(\zeta^i, -\varkappa^i) \in N((\rho^i, \bar{\beta}(\rho^i)); \operatorname{epi} \bar{\beta}).$$

Arguing as in writing out an auxiliary problem (5.8), we get that, if $\pi^{i,k} \in \mathcal{D}$, $k = 1, 2, \ldots$, is a m.a.s. of problem $(\bar{P}_{\rho^i})$ in the sense of (1.2), then $(\pi^{i,k}, \rho^i)$, $k = 1, 2, \ldots$, is a m.a.s. of the problem

$$\varkappa^i I_0(\pi) - \zeta^i \rho' + \omega|\rho' - \rho^i| \to \inf, \tag{6.3}$$
$$I_1(\pi) \in \mathcal{M} + q + \rho'\tilde{q}, \ \ \rho' \in (-\omega, \omega), \ \ \pi \in \mathcal{D},$$

where $\omega > 0$ is large enough that $\rho^i \in (-\omega, \omega)$, $i = 1, 2, \ldots$ Writing necessary conditions for m.a.s. in problem (6.3), and, then, going to the limit as $k \to \infty$, we obtain m.a.s. $\pi^i \in \mathcal{D}$, $i = 1, 2, \ldots$, for problem (P_q). This m.a.s. satisfies all relations of Theorem 2.1, and $\lambda_0^i \to \varkappa > 0$, $i \to \infty$. It follows that problem (P_q) is regular.

The proof of Theorem 2.8. Indeed, it is not hard to see that the function $\tilde{\beta}(t) \equiv \beta(q+t\xi)$ decrease monotonically on the ray $t \geq 0$. Hence, according to a classic result of functions theory of real variable, $\tilde{\beta}$ is almost everywhere differentiable for $t \geq 0$, in the classic sense. Therefore, $\hat{\partial}\tilde{\beta}(t) \neq \emptyset$ for a.e. $t \geq 0$. It follows that for a.e. $t \geq 0$ there exists a non–zero Frechet normal $(\zeta, -\varkappa) \in N((t, \tilde{\beta}(t)); \mathrm{epi}\,\tilde{\beta})$, $\varkappa > 0$. By virtue of the basic normal cone definition, there exist sequences of numbers $t^i, \zeta^i, \varkappa^i, i = 1, 2, \ldots$, such that

$$t^i \to t, \ \ \tilde{\beta}(t^i) \to \tilde{\beta}(t), (\zeta^i, -\varkappa^i) \to (\zeta, -\varkappa) \neq 0, \ \ i \to \infty, \ \ \varkappa > 0,$$
$$(\zeta^i, -\varkappa^i) \in N((t^i, \tilde{\beta}(t^i)); \mathrm{epi}\,\tilde{\beta}).$$

As in the proof of Theorem 2.7, this circumstance give us that, if $\pi^{i,k} \in \mathcal{D}$, $k = 1, 2, \ldots$, is a m.a.s. of problem $(P_{q+t^i\xi})$ in the sense of (1.2), then $(\pi^{i,k}, t^i)$, $k = 1, 2, \ldots$, is a m.a.s. of the problem

$$\varkappa^i I_0(\pi) - \zeta^i t' + \omega|t' - t^i| \to \inf, \tag{6.4}$$
$$I_1(\pi) \in \mathcal{M} + q + \rho'\tilde{q}, \ \ t' \in (-\omega + t, \omega + t), \ \ \pi \in \mathcal{D},$$

where $\omega > 0$ is large enough that $t^i \in (-\omega + t, \omega + t)$, $i = 1, 2, \ldots$ Writing necessary conditions for m.a.s. in problem (6.4), and, then, going to the limit as $k \to \infty$, we obtain m.a.s. $\pi^i \in \mathcal{D}$, $i = 1, 2, \ldots$, for problem $(P_{q+t\xi})$. This m.a.s. satisfies all relations of Theorem 2.1, and $\lambda_0^i \to \varkappa > 0$, $i \to \infty$. It follows that problem $(P_{q+t\xi})$ is regular for a.e. $t \geq 0$.

References

[1] Novozhenov, M.M.; Plotnikov, V.I. Generalized Lagrange Multiplyers Rule for Distributed Systems with State Constraints, *Diff. Uravn.;* 1982, vol.18, No.4. pp.584–692. [in Russian]

[2] Mackenroth, U. Convex Parabolic Boundary Control Problems with Pointwise State Constraints, *J. Math. Anal. Appl.*; 1982, vol.87, pp.256–277.

[3] Mackenroth, U. On Some Elliptic Optimal Control Problems with State Constraints, *Optimization;* 1986, vol.17, pp.595–607.

[4] Bergounioux, M. A Penalization Method for Optimal Control of Elliptic Problems with State Constraints, *SIAM J. Control Optim.;* 1992, vol.30, No.2, pp.305–323.

[5] Casas, E. Boundary Control of Semilinear Elliptic Equations with Pointwise State Constraints, *SIAM J. Control Optim.;* 1993, vol.31, No.4, pp.993–1006.

[6] Li, X.; Yong, J. *Optimal Control Theory for Infinite Dimensional Systems.* Birkhäuser Verlag, Basel, 1995.

[7] Bonnans, J.F.; Casas, E. An Extension of Pontryagin's Principle for State-Constrained Optimal Control of Semilinear Elliptic Equations and Variational Inequalities, *SIAM J. Control Optim.*; 1995, vol.33, No.1, pp.274–298.

[8] Casas, E. Pontryagin's Principle for State-Constrained Boundary Control Problems of Semilinear Parabolic Equations, *SIAM J. Control Optim.*; 1997, vol.35, No.4, pp.1297–1327.

[9] Raymond, J.-P.; Zidani, H. Pontryagin's Principle for State-Constrained Control Problems Governed by Parabolic Equations with Unbounded Controls, *SIAM J. Control Optim.*; 1998, vol.36, No.6, pp.1853–1879.

[10] Casas, E.; Raymond, J.-P.; Zidani, H. Pontryagin's Principle for Local Solutions of Control Problems with Mixed Control-State Constraints, *SIAM J. Control Optim.*; 2000, vol.39, No.4, pp.1182–1208.

[11] Mordukhovich, B.S.; Raymond, J.-P. Dirichlet Boundary Control of Hyperbolic Equations in the Presence of State Constraints, *Appl. Math. Optim.*; 2004, vol.49, pp.145–157.

[12] Mordukhovich, B.S.; Raymond, J.-P. Neumann Boundary Control of Hyperbolic Equations with Pointwise State Constraints, *SIAM J. Control Optim.*; 2005, vol.43, No.4, pp.1354–1372.

[13] Sumin, M.I. Suboptimal Control of Semilinear Elliptic Equations with State Constraints, I: Maximum Principle for Minimizing Sequences, Normality, *Izv. vuzov. Matematika;* 2000, No.6, pp.33–51. [in Russian]

[14] Sumin, M.I. Suboptimal Control of Semilinear Elliptic Equations with State Constraints, II: Sensitivity, Typicalness of Regular Maximum Principle, *Izv. vuzov. Matematika;* 2000, No.8, pp.52–63. [in Russian]

[15] Sumin, M.I. Optimal Control of Semilinear Elliptic Equation with State Constraint: Maximum Principle for Minimizing Sequence, Regularity, Normality, Sensitivity, *Control and Cybernetics;* 2000, vol.29, No.2, pp.449–472.

[16] Sumin, M.I. Suboptimal Control of Semilinear Elliptic Equation with Phase Constraint and with Boundary Control, *Differential Equations;* 2001, vol.37, No.2. pp.260–275.

[17] Gavrilov, V.S.; Sumin, M.I. Parametric Optimization of Nonlinear Goursat–Darboux Systems with State Constraints, *Comp. Math. Math. Phys.;* 2004, vol.44, No.6, pp.1002–1022.

[18] Gavrilov, V.S.; Sumin, M.I. Parametric Suboptimal Control Problem of Goursat–Darboux Systems with State Constraint, *Izv. vuzov. Matematika;* 2005, No.6, pp.40–52. [in Russian]

[19] Sumin, M.I. Parametric Suboptimal Control of Distributed Systems with Pointwise State Constraints, *Nonlinear Studies;* 2006. vol.13, No.3. pp. 239–249.

[20] Lions, J.-L. *Contróle des Systems Distribues Singuliers;* Bordas, Paris, 1983; Nauka, Moscow, 1987.

[21] Warga, J. *Optimal Control of Differential and Functional Equations;* Academic Press, New York, 1972.

[22] Sumin, M.I. Candidate's Dissertation in Mathematics and Physics (Gor'kov. Gos. Univ., Gor'kii, 1983). [in Russian]

[23] Sumin, M.I. On First Variation in Optimal Control Theory for Distributed Parameters Systems, *Diff. Uravn.;* 1991, vol.27, No.12. pp.2179–2181. [in Russian]

[24] Sumin, M.I. Mathematilal Theory of Suboptimal Control for Distrubuted Systems, *Doctoral Dissertation in Mathematics and Physics;* Nizhegorodskii Gos. Univ., Nizhnii Novgorod, 2000 [in Russian].

[25] Sumin, M.I. First Variation and Pontryagin Maximum Principle in Optimal Control for Partial Differential Equations, *Computational Mathematics and Mathematical Physics,*; 2009, vol.49, No.6, pp.998–1020.

[26] Smirnov, V.I. *A Course of Higher Mathematics*; Nauka, Moscow, 1959; Addison-Wesley, Reading, Mass., 1964, Vol. 5.

[27] Stane, E.M. *Singular Integrals and Differentiability Properties of Functions*; Princeton Univ. Press, Princeton, 1970.

[28] Sumin, M.I. On Minimizing Sequences in Optimal Control Problems with Constrained State Coordinates, *Diff. Uravn.;* 1986, vol.22, No.10. pp.1179–1731. [in Russian]

[29] Clarke, F. *Optimization and Nonsmooth Analysis*; Wiley, New York, 1983.

[30] Mordukhovich, B.S. *Variational Analysis and Generalized Differentiation, I: Basic Theory*; Springer, Berlin, 2006.

[31] Mordukhovich, B.S. *Variational Analysis and Generalized Differentiation, II: Applications*; Springer, Berlin, 2006.

[32] Mordukhovich, B.S.; Shao, Y. Nonsmooth sequential analysis in asplund spaces, *Trans. Amer. Math. Soc.;* 1996, vol.346, No.4, pp.1235–1280.

[33] Ekeland, I. On the Variational Principle, *J. Math. Anal. Appl.*; 1974, vol. 47, pp. 324–353.

[34] Gavrilov, V.S. An Unique Existence of the Solution to a Nonhomogeneous Third Boundary–Value Problem for a Divergence Form Semilinear Hyperbolic Partial Differential Equation, *Vestnik Nizhegorodskogo universiteta im. N.I. Lobachevskogo;* 2009, No.1, pp.104–111. [in Russian]

[35] Ladyzhenskaya, O.A. *Boundary–Value Problems of Mathematical Physic;* Nauka, Moscow, 1973. [in Russian]

[36] Gavrilov, V.S. An Unique Existence of the Solution to a Nonhomogeneous Third Boundary–Value Problem for a Divergence form Linear Hyperbolic Partial Differential Equation with Radon Measure in the Right–Hand Side Part, *Vestnik Nizhegorodskogo universiteta im. N.I. Lobachevskogo;* 2009, No.5, pp.158–162. [in Russian]

[37] Osipov, Yu.S.; Vasil'ev, F.P.; Potapov, M.M. *Fundamentals of the Dynamic Regularization Method*; Mosk. Gos. Univ., Moscow, 1999 [in Russian].

[38] Sumin, M.I. Suboptimal Control for Distributed Parameters Systems: Minimizing Sequences and the Value Function, *Comp. Math. Math. Phys.;* 1997, vol.37, No.1, pp.21–39.

[39] Ward, A.L. Differentiability of Vector Monotone Functions *Proc. London Math. Soc.;* 1935, vol. 32, No.2, pp. 339–362.

[40] Plotnikov, V.I. Sufficient Optimality Conditions for Controlled Systems of General Form, *Inform. mat.;* 1970, vol.5(42), Akad. Nauk SSSR. Nauchnyi Sovet po Kompleksnoj probleme "Kibernetika", Moscow, pp.38–46. [in Russian]

[41] Plotnikov, V.I.; Sumin, M.I. On Conditions on the Elements of Minimizing Sequences of Optimal Control Problems, *Soviet Math. Dokl.;* 1985, vol.31, No.1, pp.73–77.

[42] Sumin, M.I. On Sufficient Conditions on Elements of Minimizing Sequences in Optimal Control Problems, *Zh. Vychisl. Mat. Mat. Fiz.;* 1985, vol.25, No.1, pp.23–31. [in Russian]

In: Control Theory and Its Applications   ISBN 978-1-61668-384-9
Editor: Vito G. Massari, pp. 145-165   © 2011 Nova Science Publishers, Inc.

Chapter 5

ROBUST ADAPTIVE SLIDING MODE CONTROL LAW FOR INDUCTION MOTORS

*Oscar Barambones**
Dpto. Ingeniería de Sistemas y Automática
Universidad del País Vasco.
EUI de Vitoria. Nieves Cano 12.

Abstract

A novel sensorless adaptive robust control law is proposed to improve the trajectory tracking performance of induction motors. The proposed design employs the so called vector (or field oriented) control theory for the induction motor drives and the designed control law is based on an integral sliding-mode algorithm that overcomes the system uncertainties. The proposed sliding-mode control law incorporates an adaptive switching gain to avoid calculating an upper limit of the system uncertainties. The proposed design also includes a new method in order to estimate the rotor speed. In this method, the rotor speed estimation error is presented as a first order simple function based on the difference between the real stator currents and the estimated stator currents.

The stability analysis of the proposed controller under parameter uncertainties and load disturbances is provided using the Lyapunov stability theory. Finally simulated results show, on the one hand that the proposed controller with the proposed rotor speed estimator provides high-performance dynamic characteristics, and on the other hand that this

*E-mail address: oscar.barambones@ehu.es. Tel: +34 945013235; Fax: +34 945013270

scheme is robust with respect to plant parameter variations and external load disturbances.

1. Introduction

Field oriented control method is widely used for advanced control of induction motor drives. By providing decoupling of torque and flux control demands, the vector control can govern an induction motor drive similar to a separate excited direct current motor without sacrificing the quality of the dynamic performance.

However, the field oriented control of induction motor drives presents two main problems that have been providing quite a bit research interest in the last decade. The first one relies on the uncertainties in the machine models and load torque, and the second one is the precise computation of the motor speed without using speed sensors.

The decoupling characteristics of the vector control is sensitive to machine parameters variations. Moreover, the machine parameters and load characteristics are not exactly known, and may vary during motor operations.

To overcome the above system uncertainties, the variable structure control strategy using the sliding-mode has been focussed on many studies and research for the control of the AC servo drive system in the past decade [1], [2], [5], [8], [11].

However the traditional sliding control schemes requires the prior knowledge of an upper bound for the system uncertainties since this bound is employed in the switching gain calculation. This upper bound should be determined as precisely as possible, because as higher is the upper bound higher value should be considered for the sliding gain, and therefore the control effort will also be high, which is undesirable in a practice. In order to surmount this drawback, in the present paper it is proposed an adaptive law to calculate the sliding gain which avoids the necessity of calculate an upper bound of the system uncertainties.

Otherwise, a suitable speed control of an induction motor requires a precise speed information, therefore, a speed sensor, such a resolver and encoder, is usually adhered to the shaft of the motor to measure the motor speed. However, a speed sensor can not be mounted in some cases, such as motor drives in a adverse environments, or high-speed motor drives. Moreover, such sensors lower the system reliability and require special attention to noise. Therefore, sensorless induction motor drives are widely used in industry for their reliability and flexibility, particularly in hostile environments. Speed estimation methods using Model Reference Adaptive System MRAS are the most commonly used as they are easy to design and implement [3], [7], [14]. However, the performance of these methods is deteriorated at low speed because of the increment of nonlinear characteristics [6], [9].

In this paper the authors proposes a robust sensorless vector control scheme consisting on the one hand of an adaptive rotor speed estimation method based on MRAS in order to improve the performance of a sensorless vector controller in a low speed region. In the presented method, the stator current is estimated and then it is compared with the real stator current. Next, the stator current error is used in the adaptive law to estimate the rotor speed. The proposed method can provide a fast speed estimation and improve the performance of other speed estimation methods in a low speed region and at zero-speed. In addition, the proposed variable structure control algorithm presents an adaptive sliding gain that is estimated on-line in order to compensate the present system uncertainties. Using the proposed sensorless variable structure control to govern the induction motor drive, the rotor speed becomes insensitive to variations in the motor parameters and load disturbances. Moreover, the proposed control scheme, provides a good transient response and exponential convergence of the speed trajectory tracking in spite of parameter uncertainties and load torque disturbances.

This report is organized as follows. The rotor speed estimation is introduced in Section 2. In section 3, the proposed robust speed control with adaptative sliding gain is presented, and in section 4 it is proposed a continuous approximation of the control law. Then the closed loop stability of the proposed scheme is demonstrated using the Lyapunov stability theory, and the exponential convergence of the controlled speed is also provided. In the Section 5, some simulation results are presented. Finally some concluding remarks are stated in the last Section.

2. Proposed Rotor Speed Estimator

Many schemes [10], [12], based on simplified motor models have been devised to sense the speed of the induction motor from measured terminal quantities for control purposes. In order to obtain an accurate dynamic representation of the motor speed, it is necessary to base the calculation on the coupled circuit equations of the motor. However, the performance of these methods is deteriorated at a low speed because of the increment of nonlinear characteristic of the system [9].

The current paper proposes a new rotor speed estimation method to improve the performance of a sensorless vector controller in the low speed region and at zero speed.

Since the motor voltages and currents are measured in a stationary frame of reference, it is also convenient to express these equations in that stationary frame.

From the stator voltage equations in the stationary frame it is obtained [4]:

$$\dot{\psi}_{dr} = \frac{L_r}{L_m}\left[v_{ds} - R_s i_{ds} - \sigma L_s \frac{d}{dt}i_{ds}\right] \tag{1}$$

$$\dot{\psi}_{qr} = \frac{L_r}{L_m}\left[v_{qs} - R_s i_{qs} - \sigma L_s \frac{d}{dt}i_{qs}\right] \tag{2}$$

where ψ is the flux linkage; L is the inductance; v is the voltage; R is the resistance; i is the current and $\sigma = 1 - L_m^2/(L_r L_s)$ is the motor leakage coefficient. The subscripts r and s denotes the rotor and stator values respectively refereed to the stator, and the subscripts d and q denote the dq-axis components in the stationary reference frame.

Using the rotor flux and motor speed, the stator current is represented as:

$$i_{ds} = \frac{1}{L_m}\left[\psi_{dr} + w_r T_r \psi_{qr} + T_r \dot{\psi}_{dr}\right] \tag{3}$$

$$i_{qs} = \frac{1}{L_m}\left[\psi_{qr} - w_r T_r \psi_{dr} + T_r \dot{\psi}_{qr}\right] \tag{4}$$

where w_r is the rotor electrical speed and $T_r = L_r/R_r$ is the rotor time constant.

From the equations (3) and (4) and using the estimated speed, the stator current is estimated as:

$$\hat{i}_{ds} = \frac{1}{L_m}\left[\psi_{dr} + \hat{w}_r T_r \psi_{qr} + T_r \dot{\psi}_{dr}\right] \tag{5}$$

$$\hat{i}_{qs} = \frac{1}{L_m}\left[\psi_{qr} - \hat{w}_r T_r \psi_{dr} + T_r \dot{\psi}_{qr}\right] \tag{6}$$

where $\hat{i}_{ds}$ and $\hat{i}_{qs}$ are the estimated stator currents and $\hat{w}_r$ is the estimated rotor electrical speed.

Subtracting the equations of the estimated stator currents (5) and (6) from the equations of the stator currents (3) and (4) the difference in the stator current is obtained as:

$$i_{ds} - \hat{i}_{ds} = \frac{T_r}{L_m} \psi_{qr} (w_r - \hat{w}_r) \tag{7}$$

$$i_{qs} - \hat{i}_{qs} = -\frac{T_r}{L_m} \psi_{dr} (w_r - \hat{w}_r) \tag{8}$$

In the above equations (7) and (8), the difference of the stator current and the estimated stator current is a sinusoidal value because it is a function of the rotor flux. However, if equation (7) is multiplied by ψ_{qr} and equation (8) is multiplied by ψ_{dr} and then are added together it is obtained:

$$(i_{ds} - \hat{i}_{ds})\psi_{qr} - (i_{qs} - \hat{i}_{qs})\psi_{dr} =$$
$$\frac{T_r}{L_m} \psi_{qr} (w_r - \hat{w}_r) (\psi_{dr}^2 + \psi_{dr}^2) \tag{9}$$

Unlike the equations (7) and (8), equation (9) uses the rotor flux magnitude which remains constant. From equation (9) the error of the rotor speed is obtained as follows:

$$e_{w_r} = w_r - \hat{w}_r = c \left[(i_{ds} - \hat{i}_{ds})\psi_{qr} - (i_{qs} - \hat{i}_{qs})\psi_{dr} \right] \tag{10}$$

where:

$$c = \frac{L_m}{T_r} \frac{1}{\psi_{dr}^2 + \psi_{qr}^2} = \frac{L_m}{T_r \psi_r^2}$$

Therefore, from the equation (10) the speed estimation error is calculated from the stator current and rotor flux.

Using Lyapunov stability theory we can derive the following adaptation law for speed estimation:

$$\frac{d\hat{w}_r}{dt} = \alpha e_{w_r}, \quad \alpha > 0 \tag{11}$$

where α is de adaptation gain that should be chosen greater than zero.

To demonstrate that the previous adaptation law makes the estimated speed error drops into zero, we can define the following Lyapunov function candidate:

$$V(t) = \frac{1}{2} e_{w_r}^2(t)$$

On the basis of the fact that the velocity of outer control loop is much slower than the estimated inner loop, hence the assumption of w_r approaching a constant is reasonable on deriving the following equations. Then, the time derivative of the previous Lyapunov function candidate is:

$$
\begin{aligned}
\dot{V}(t) &= e_{w_r} \dot{e}_{w_r} \\
&= e_{w_r}(-\dot{\hat{w}}_r) \\
&= -\alpha\, e_{w_r}^2
\end{aligned}
\tag{12}
$$

Using the Lyapunov's direct method, since $V(t)$ is clearly positive-definite, $\dot{V}(t)$ is negative definite and $V(t)$ tends to infinity as $e_{w_r}(t)$ tends to infinity, then the equilibrium at the origin $e_{w_r}(t) = 0$ is globally asymptotically stable. Therefore $e_{w_r}(t)$ tends to zero as the time t tends to infinity.

Moreover, taking into account the previous Lyapunov function we can conclude that the rotor speed error converges to zero exponentially. From equation (12) we can obtain that V derivative verifies:

$$
\dot{V}(t) = -\alpha\, e_{w_r}^2 = -\frac{\alpha}{2} V(t)
\tag{13}
$$

The solution of the previous differential equation is:

$$
V(t) = \frac{1}{2} e_{w_r}^2(t) = V(t_0) \exp(-\frac{\alpha}{2} t)
$$

which implies that the rotor speed error converges to zero exponentially.

Therefore, the rotor speed w_r can be calculated using the proposed speed estimator which only make use of the measured stator voltages and currents in order to estimate the rotor speed.

3. Variable Structure Robust Speed Control with Adaptive Sliding Gain

In general, the mechanical equation of an induction motor can be written as:

$$
J\dot{w}_m + Bw_m + T_L = T_e
\tag{14}
$$

where J and B are the inertia constant and the viscous friction coefficient of the induction motor system respectively; T_L is the external load; w_m is the rotor mechanical speed in angular frequency, which is related to the rotor electrical speed by $w_m = 2w_r/p$ where p is the pole numbers and T_e denotes the generated torque of an induction motor, defined as [4]:

$$T_e = \frac{3p}{4}\frac{L_m}{L_r}(\psi_{dr}^e i_{qs}^e - \psi_{qr}^e i_{ds}^e) \tag{15}$$

where ψ_{dr}^e and ψ_{qr}^e are the rotor-flux linkages, with the subscript 'e' denoting that the quantity is refereed to the synchronously rotating reference frame; i_{qs}^e and i_{ds}^e are the stator currents, and p is the pole numbers.

The relation between the synchronously rotating reference frame and the stationary reference frame is performed by the so-called reverse Park's transformation:

$$\begin{bmatrix} x_a \\ x_b \\ x_c \end{bmatrix} = \begin{bmatrix} \cos(\theta_e) & -\sin(\theta_e) \\ \cos(\theta_e - 2\pi/3) & -\sin(\theta_e - 2\pi/3) \\ \cos(\theta_e + 2\pi/3) & -\sin(\theta_e + 2\pi/3) \end{bmatrix} \begin{bmatrix} x_d \\ x_q \end{bmatrix} \tag{16}$$

where θ_e is the angle position between the d-axix of the synchronously rotating and the stationary reference frames, and it is assumed that the quantities are balanced.

The angular position of the rotor flux vector ($\bar{\psi}_r$) related to the d-axis of the stationary reference frame may be calculated by means of the rotor flux components in this reference frame (ψ_{dr}, ψ_{qr}) as follows:

$$\theta_e = \arctan\left(\frac{\psi_{qr}}{\psi_{dr}}\right) \tag{17}$$

where θ_e is the angular position of the rotor flux vector.

Using the field-orientation control principle [4] the current component i_{ds}^e is aligned in the direction of the rotor flux vector $\bar{\psi}_r$, and the current component i_{qs}^e is aligned in the direction perpendicular to it. At this condition, it is satisfied that:

$$\psi_{qr}^e = 0, \qquad \psi_{dr}^e = |\bar{\psi}_r| \tag{18}$$

Therefore, taking into account the previous results, the equation of induction motor torque (15) is simplified to:

$$T_e = \frac{3p}{4}\frac{L_m}{L_r}\psi_{dr}^e i_{qs}^e = K_T i_{qs}^e \tag{19}$$

where K_T is the torque constant, and is defined as follows:

$$K_T = \frac{3p}{4} \frac{L_m}{L_r} \psi_{dr}^{e*} \qquad (20)$$

where ψ_{dr}^{e*} denotes the command rotor flux.

With the above mentioned proper field orientation, the dynamic of the rotor flux is given by [4]:

$$\frac{d\psi_{dr}^e}{dt} + \frac{\psi_{dr}^e}{T_r} = \frac{L_m}{T_r} i_{ds}^e \qquad (21)$$

Then, the mechanical equation (14) becomes:

$$\dot{w}_m + a\, w_m + f = b\, i_{qs}^e \qquad (22)$$

where the parameters are defined as:

$$a = \frac{B}{J}, \quad b = \frac{K_T}{J}, \quad f = \frac{T_L}{J}; \qquad (23)$$

Now, we are going to consider the previous mechanical equation (22) with uncertainties as follows:

$$\dot{w}_m = -(a + \triangle a)w_m - (f + \triangle f) + (b + \triangle b)i_{sq} \qquad (24)$$

where the terms $\triangle a$, $\triangle b$ and $\triangle f$ represents the uncertainties of the terms a, b and f respectively. It should be noted that these uncertainties are unknown, and that the precise calculation of its upper bound are, in general, rather difficult to achieve.

Let us define the tracking speed error as follows:

$$e(t) = w_m(t) - w_m^*(t) \qquad (25)$$

where w_m^* is the rotor speed command.

Taking the derivative of the previous equation with respect to time yields:

$$\dot{e}(t) = \dot{w}_m - \dot{w}_m^* = -a\, e(t) + u(t) + d(t) \qquad (26)$$

where the following terms have been collected in the signal $u(t)$,

$$u(t) = b\, i_{sq}(t) - a\, w_m^*(t) - f(t) - \dot{w}_m^*(t) \qquad (27)$$

and the uncertainty terms have been collected in the signal $d(t)$,

$$d(t) = -\triangle a\, w_m(t) - \triangle f(t) + \triangle b\, i_{sq}(t) \tag{28}$$

To compensate for the above described uncertainties that are presented in the system, it is proposed a sliding adaptive control scheme. In the sliding control theory, the switching gain must be constructed so as to attain the sliding condition (Utkin 1993). In order to meet this condition a suitable choice of the sliding gain should be made to compensate for the uncertainties. For selecting the sliding gain vector, an upper bound of the parameter variations, unmodelled dynamics, noise magnitudes, etc. should be known, but in practical applications there are situations in which these bounds are unknown, or at least difficult to calculate. A solution could be to choose a sufficiently high value for the sliding gain, but this approach could cause a to high control signal, or at least more activity control than it is necessary in order to achieve the control objective.

One possible way to overcome this difficulty is to estimate the gain and to update it by some adaptation law, so that the sliding condition is achieved.

Now, we are going to propose the sliding variable $S(t)$ with an integral component as:

$$S(t) = e(t) + \int_0^t (a+k)e(\tau)\, d\tau \tag{29}$$

where k is a constant gain, and a is a parameter that was already defined in equation (23).

Then the sliding surface is defined as:

$$S(t) = e(t) + \int_0^t (a+k)e(\tau)\, d\tau = 0 \tag{30}$$

Now, we are going to design a variable structure speed controller, that incorporates an adaptive sliding gain, in order to control the AC motor drive.

$$u(t) = -k\, e(t) - \hat{\beta}(t)\gamma\, \mathrm{sgn}(S) \tag{31}$$

where the k is the gain defined previously, $\hat{\beta}$ is the estimated switching gain, γ is a positive constant, S is the sliding variable defined in eqn. (29) and $\mathrm{sgn}(\cdot)$ is the signum function.

The switching gain $\hat{\beta}$ is adapted according to the following updating law:

$$\dot{\hat{\beta}} = \gamma\, |S| \qquad \hat{\beta}(0) = 0 \tag{32}$$

where γ is a positive constant that let us choose the adaptation speed for the sliding gain.

In order to obtain the speed trajectory tracking, the following assumptions should be formulated:

($\mathcal{A}$1) The gain k must be chosen so that the term $(a + k)$ is strictly positive. Therefore the constant k should be $k > -a$.

($\mathcal{A}$2) There exits an unknown finite non-negative switching gain β such that

$$\beta > d_{max} + \eta \qquad \eta > 0$$

where $d_{max} \geq |d(t)|$ $\forall t$ and η is a positive constant.

Note that this condition only implies that the uncertainties of the system are bounded magnitudes.

($\mathcal{A}$3) The constant γ must be chosen so that $\gamma > 0$.

Theorem 1 *Consider the induction motor given by equation (24). Then, if assumptions ($\mathcal{A}$1), ($\mathcal{A}$2) and ($\mathcal{A}$3) are verified, the control law (31) leads the rotor mechanical speed $w_m(t)$ so that the speed tracking error $e(t) = w_m(t) - w_m^*(t)$ tends to zero as the time tends to infinity.*

The proof of this theorem will be carried out using the Lyapunov stability theory (Barambones 2007).

Finally, the torque current command, $i_{sq}^*(t)$, can be obtained directly substituting eqn. (31) in eqn. (27):

$$i_{sq}^*(t) = \frac{1}{b}\left[k\,e - \hat{\beta}\gamma\,\mathrm{sgn}(S) + a\,w_m^* + \dot{w}_m^* + f\right] \tag{33}$$

Therefore, the proposed variable structure speed control with adaptive sliding gain resolves the speed tracking problem for the induction motor, with some uncertainties in mechanical parameters and load torque.

4. Continuous Approximation of Switching Control Law

A frequently encountered problem in sliding control is that the control signal given by eqn.(31) is not smooth since the sliding control law is discontinuous

across the sliding surfaces, which causes the chattering phenomenon. Chattering is undesirable in practice, since it involves high control activity and further
may excite high-frequency dynamics. This situation can be avoided by smoothing out the control chattering within a thin boundary layer of thickness $\xi > 0$
neighboring the switching surface (Slotine 1991, Barambones 2002). On the
other hand, it is well known that when in an adaptive control system the signals
are not persistently exciting the parameter drift phenomenon may appear (Slotine and Li 1991). In these situations many different strategies can be applied,
from complex methods to confer self-excitation capability to the system without
the presence of external exciting signals, to simpler approaches based on the use
of dead zones, as it will be done in this paper to avoid the possible parameter
drift phenomenon that may appear in the proposed sliding gain adaptation law
(eqn. 32)

In this way, some modifications should be done in the control law (31) and
in the sliding gain adaptation law (32), to overcome the above mentioned problems:

i) In order to smooth the control law (31), the sign function included in it is
replaced by a saturation function, so that it becomes:

$$u(t) = -k\,e(t) - \hat{\beta}(t)\gamma\,\mathrm{sat}\left(\frac{S}{\xi}\right) \qquad (34)$$

where the saturation function $sat(\cdot)$ is defined in the usual way:

$$\mathrm{sat}\left(\frac{S}{\xi}\right) = \begin{cases} \mathrm{sgn}(S) & if \quad |S| > \xi \\ \dfrac{S}{\xi} & otherwise. \end{cases}$$

and ξ represents the thickness of the boundary layer neighboring the
switching surface.

ii) In order to avoid the parameter drift phenomenon, the sliding gain adaptation law is modified to:

$$\dot{\hat{\beta}} = \gamma\,|S_o| \qquad\qquad \hat{\beta}(0) = 0 \qquad (35)$$

where S_o is defined by:

$$S_o = S - \xi\,\mathrm{sat}\left(\frac{S}{\xi}\right)$$

It is interesting to point out that S_o is a measure of the distance from the sliding surface S to the interval $[-\xi, \xi]$:

$$
S_o = \begin{cases} S - \xi & if \quad |S| > \xi \\ 0 & otherwise. \end{cases} \tag{36}
$$

From the previous equation it is concluded that $\dot{S}_o = \dot{S}$ when S is outside the interval $[-\xi, \xi]$, while $\dot{S}_o = 0$ otherwise.

Theorem 2 *Consider the induction motor given by equation (24). Then, if assumptions $(\mathcal{A}1)$, $(\mathcal{A}2)$ and $(\mathcal{A}3)$ are verified, the control law (34) leads the rotor mechanical speed $w_m(t)$ so that the speed tracking error $e(t) = w_m(t) - w_m^*(t)$ can be made as small as desired by choosing an adequately small boundary layer thickness ξ.*

The proof of this theorem will be carried out using the Lyapunov stability theory.
Proof : Let us define the following Lyapunov function candidate:

$$
V(t) = \frac{1}{2} S_o(t) S_o(t) + \frac{1}{2} \tilde{\beta}(t) \tilde{\beta}(t) \tag{37}
$$

whose time derivative is given by:

$$
\begin{aligned}
\dot{V}(t) = & \; S_o(t)\dot{S}(t) + \tilde{\beta}(t)\dot{\tilde{\beta}}(t) \\
= & \; S_o \cdot [\dot{e} + (a+k)e] + \tilde{\beta}(t)\dot{\tilde{\beta}}(t) \\
= & \; S_o \cdot [(-a\,e + u + d) + (k\,e + a\,e)] + \tilde{\beta}\,\gamma|S_o| \\
= & \; S_o \cdot [u + d + k\,e] + (\hat{\beta} - \beta)\gamma|S_o| \\
= & \; S_o \cdot \left[-k\,e - \hat{\beta}\gamma\,\mathrm{sat}(S/\xi) + d + k\,e\right] + (\hat{\beta} - \beta)\gamma|S_o| \\
= & \; S_o \cdot \left[d - \hat{\beta}\gamma\,\mathrm{sat}(S/\xi)\right] + \hat{\beta}\gamma|S_o| - \beta\gamma|S_o| \\
= & \; d\,S_o - \hat{\beta}\gamma|S_o| + \hat{\beta}\gamma|S_o| - \beta\gamma|S_o| \\
\leq & \; |d||S_o| - \beta\gamma|S| \\
\leq & \; |d||S_o| - (d_{max} + \eta)\gamma|S| \\
= & \; |d||S_o| - d_{max}\,\gamma|S_o| - \eta\,\gamma|S_o| \\
\leq & \; -\eta\,\gamma|S_o|
\end{aligned} \tag{38}
$$

$$
\tag{39}
$$

then

$$\dot{V}(t) \leq 0 \tag{40}$$

It should be noted that in the proof the equations (29), (26), (34) and (35) and the assumptions $(\mathcal{A}\,2)$ and $(\mathcal{A}\,3)$ have been used. It has been also used that by means of S_o definition (eqn. 36), it is obtained that $S_o \operatorname{sat}(S/\xi) = |S_o|$.

Using the Lyapunov's direct method, since $V(t)$ is clearly positive-definite, $\dot{V}(t)$ is negative semidefinite and $V(t)$ tends to infinity as $S_o(t)$ and $\tilde{\beta}(t)$ tends to infinity, then the equilibrium at the origin $[S_o(t), \tilde{\beta}(t)] = [0,0]$ is globally stable, and therefore the variables $S_o(t)$ and $\tilde{\beta}(t)$ are bounded. Since $S_o(t)$ is bounded then $S(t)$ is also bounded, and hence it is deduced that $e(t)$ is bounded. From equations (26) and (30) it is obtained that

$$\dot{S}(t) = ke(t) + d(t) + u(t) \tag{41}$$

Then, from equation (41) we can conclude that $\dot{S}(t)$ is bounded because $e(t)$, $u(t)$ and $d(t)$ are bounded. Since $\dot{S}(t)$ is bounded then from equation (36) it may be deduced that $\dot{S}_o(t)$ is a bounded value.
Now, from equation (38) it is concluded that

$$\ddot{V}(t) = d\,\dot{S}_o - \beta\gamma\frac{d}{dt}|S_o(t)| \tag{42}$$

which is a bounded value because $\dot{S}_o(t)$ is bounded.

Under these conditions, since $\ddot{V}$ is bounded, $\dot{V}$ is a uniformly continuous function, so by means of Barbalat's lemma we can conclude that $\dot{V} \to 0$ as $t \to \infty$, which implies that $S_o(t) \to 0$ as $t \to \infty$, or equivalently that S converges to the interval $[-\xi, \xi]$ asymptotically, so under the definition of S, the tracking error $(e = w_m - w_m^*)$ converges to a small value depending on the boundary thickness ξ.

The torque current command, $i_{sq}^*(t)$, can be obtained directly substituting eqn. (34) in eqn. (27):

$$i_{sq}^*(t) = \frac{1}{b}\left[k\,e - \hat{\beta}\gamma\operatorname{sat}\left(\frac{S}{\xi}\right) + a\,w_m^* + \dot{w}_m^* + f\right] \tag{43}$$

5. Simulation Results

In this section we will study the speed regulation performance of the proposed adaptive sliding-mode field oriented control under reference and load torque variations by means of simulation examples.

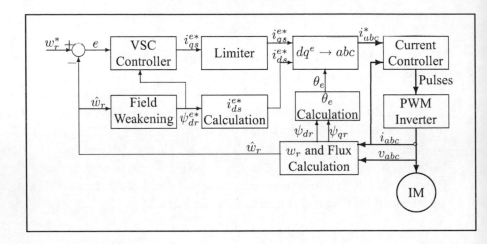

Figure 1. Block diagram of the proposed adaptive sliding-mode control.

The block diagram of the proposed robust control scheme is presented in figure 1.

The block diagram of the proposed robust control scheme is presented in figure 1, where the block 'VSC Controller' represent the proposed adaptive sliding-mode controller, and it is implemented by equations (29), (33), and (32). The block 'limiter' limits the current applied to the motor windings so that it remains within the limit value, and it is implemented by a saturation function. The block '$dq^e \rightarrow abc$' makes the conversion between the synchronously rotating and stationary reference frames, and is implemented by equation (16). The block 'Current Controller' consists of a three hysteresis-band current PWM control, which is basically an instantaneous feedback current control method of PWM where the actual current (i_{abc}) continually tracks the command current (i^*_{abc}) within a hysteresis band. The block 'PWM Inverter' is a six IGBT-diode bridge inverter with 780 V DC voltage source. The block 'Field Weakening' gives the flux command based on rotor speed, so that the PWM controller does not saturate. The block 'i^{e*}_{ds} Calculation' provides the current reference i^{e*}_{ds} from the rotor flux reference through the equation (21). The block 'w_r and Flux Calculation' represent the proposed rotor speed estimator and flux calculator, and is implemented by the equations (11), (1) and (2) respectively and the block 'IM' represents the induction motor.

The induction motor used in this case study is a 50 HP, 460 V, four pole, 60 Hz motor having the following parameters: $R_s = 0.087\,\Omega$, $R_r = 0.228\,\Omega$, $L_s = 35.5\,mH$, $L_r = 35.5\,mH$, and $L_m = 34.7\,mH$.

The system has the following mechanical parameters: $J = 1.662\,kg.m^2$ and $B = 0.12\,N.m.s$. It is assumed that there are an uncertainty around 20 % in the system parameters, that will be overcome by the proposed adaptive sliding control.

The following values have been chosen for the controller parameters: $k = 25$, $\gamma = 15$ and $\xi = 0.1$.

In the following examples the motor starts from a standstill state and we want the rotor speed to follow a speed command that starts from zero and accelerates until the rotor speed is $130\,rad/s$. The system starts with an initial load torque $T_L = 0\,N.m$, and at time $t = 1.6\,s$ the load torque steps from $T_L = 0\,N.m$ to $T_L = 200\,N.m$ and it is assumed that there is an uncertainty around 70 % in the load torque.

5.1. First Example

In this example, shown in Figures (2-4) it is employed the control law proposed in section 3 where it is used a signum function in the sliding mode control law.

Figure 2 shows the desired rotor speed (dashed line) and the real rotor speed (solid line). As it may be observed, after a transitory time in which the sliding gain is adapted, the rotor speed tracks the desired speed in spite of system uncertainties. However, at time $t = 1.6\,s$ a little speed error can be observed. This error appears because of the torque increment at this time, and then the control system lost the so called 'sliding mode' because the actual sliding gain is too small to overcome the new uncertainty introduced in the system due to the new torque. But then, after a small time the sliding gain is adapted so that this gain can compensate the system uncertainties and so the rotor speed error is eliminated.

Figure 3 presents the time evolution of the estimated sliding gain. The sliding gain starts from zero and then it is increased until its value is high enough to compensate for the system uncertainties. Then at time $0.13\,s$ the sliding gain is remained constant because the system uncertainties remain constant as well. Later at time $1.6\,s$, there is an increment in the system uncertainties caused by the rise in the load torque. Therefore the sliding gain is adapted once again in order to overcome the new system uncertainties. As it can be seen in the figure, after the sliding gain is adapted it remains constant again, since the system uncertainties remains constant as well.

It should be noted that the adaptive sliding gain allows to employ a smaller sliding gain. In this way it is not necessary to choose the slading gain value high enough to compensate all the possible system uncertainties as used in conventional sliding control laws. With the proposed adaptive scheme the sliding gain is adapted (if necessary) when a new uncertainty appears in the system in order to surmount this uncertainty.

Figure 4 shows the motor torque. This figure shows that in the initial state, the motor torque has a high initial value in the speed acceleration zone because it is necessary a high torque to increment the rotor speed owing to the rotor inertia, then the value decreases in a constant region and finally increases due to the load torque increment. In this figure it may be observed that in the motor torque appears the so-called chattering phenomenon due to the signum function presented in the control law. It should be noted that the chattering involves high control activity and may further excite high-frequency dynamics. This undesirable effect can be avoided using the modified adaptive sliding control law proposed in Section 4, as it is shown in the next simulation results.

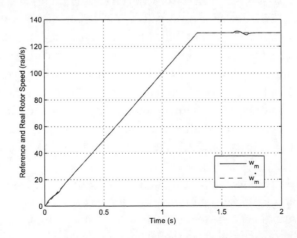

Figure 2. Reference and real rotor speed signals (rad/s).

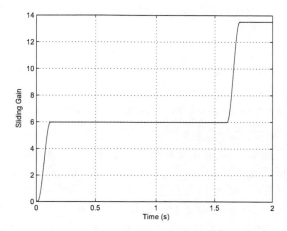

Figure 3. Estimated sliding Gain.

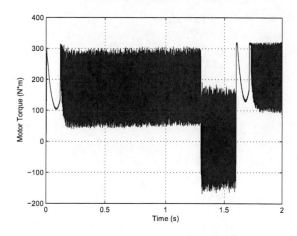

Figure 4. Motor torque (N.m).

5.2. Second Example

In this second example, shown in Figures (5-7) it is employed the control law proposed in section 4 where the control law is smoothed out within a boundary layer in order to avoid the chattering phenomenon that is undesirable in practice.

Figure 5 shows the desired rotor speed (dashed line) and the real rotor speed (solid line). Similarly to the previous example, after a transitory time in which the sliding gain is adapted, the rotor speed tracks the desired speed in spite of system uncertainties.

Figure 6 presents the time evolution of the estimated sliding gain. As before, the sliding gain starts from zero and then it is increased until its value is high enough in order to compensate the system uncertainties. Then at time $t = 1.6\,s$, the sliding gain is adapted once again in order to overcome the new system uncertainties caused by the the rise in the load torque.

Figure 7 shows the motor torque. As in the first example the motor torque presents a high initial value in the speed acceleration zone, then the value decreases in a constant region and at time $t = 1.6\,s$ the motor torque increases due to the load torque increment. However, as it may be observed, unlike the previous example, in the present example the chattering phenomenon does not appear in the motor torque, because the control signal is smoothed out within a boundary layer by means of a saturation function instead of signum function. In consequence, the smoothed control law proposed in Section 4 presents lower control activity than the previous one presented in Section 3, and therefore the motion of the motor drive in this case would be softer than in the previous case.

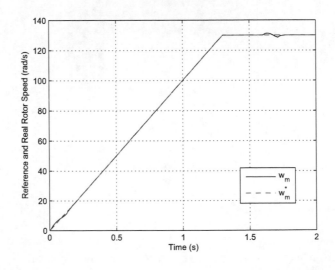

Figure 5. Reference and real rotor speed signals (rad/s).

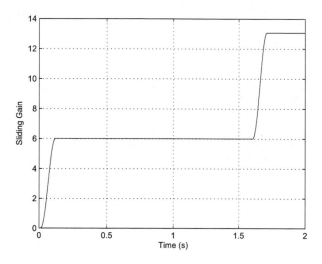

Figure 6. Estimated sliding Gain.

6. Conclusion

In this paper a new adaptive sliding mode vector control has been presented. It is proposed a variable structure control which has an integral sliding surface to relax the requirement of the acceleration signal, that is usual in conventional sliding mode speed control techniques. Due to the nature of the sliding control this control scheme is robust under uncertainties caused by parameter error or by changes in the load torque. Moreover, the proposed variable structure control incorporates an adaptive algorithm to calculate the sliding gain value. The adaptation of the sliding gain, on the one hand avoids the necessity of computing the upper bound of the system uncertainties, and on the other hand allows to employ as smaller sliding gain as possible to overcome the actual system uncertainties. Then the control signal of our proposed variable structure control scheme will be smaller than the control signals of the traditional variable structure control schemes, because in the last one the sliding gain value should be chosen high enough to overcome all the possible uncertainties that could appear in the system along the time. Also, the proposed control law is smoothed out within a boundary layer in order to avoid the chattering phenomenon since it involves high control activity and may further excite high-frequency dynamics.

The closed loop stability of the presented design has been proved thought

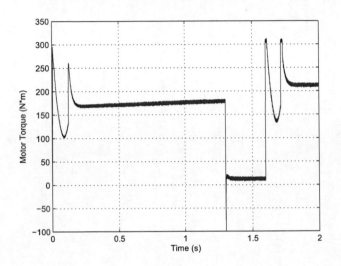

Figure 7. Motor torque (N.m).

Lyapunov stability theory. Finally, by means of simulation examples, it has been shown that the proposed control scheme performs reasonably well in practice, and that the speed tracking objective is achieved under uncertainties in the parameters and load torque.

References

[1] Barambones, O. and Garrido, A.J., 2004, A sensorless variable structure control of induction motor drives, *Electric Power Systems Research*, **72**, 21-32.

[2] Benchaib, A. and Edwards, C., 2000, Nonlinear sliding mode control of an induction motor, *Int. J. of Adaptive Control and Signal Procesing*, **14**, 201-221.

[3] Bose, B.K., 1993, Power electronics and motion control-technology status and recent trends *IEEE Trans. on Ind. Appl.*, vol.29, pp.902-909.

[4] Bose, B.K., 2001, *Modern Power Electronics and AC Drives.*, Prentice Hall, New Jersey.

[5] Chern, T.L., Chang, J. and Tsai, K.L.,1998, Integral variable structure control based adaptive speed estimator and resistance identifier for an induction motor. *Int. J. of Control*, **69**, 31-47.

[6] Kojabadi, H.M. and Chang, L., 2002, Model reference adaptive system pseudoreduced-order flux observer for very low speed and zero speed stimation in sensorless induction motor drives., *IEEE Annual Power Electronics Specialists Conference*, Australia, vol. **1**, pp. 301-308.

[7] Lehonhard, W., 1996, *Control of Electrical Drives*. Springer, Berlin.

[8] Park M.H. and Kim, K.S., 1991, Chattering reduction in the position contol of induction motor using the sliding mode, *IEEE Trans. Power Electron.*, **6** 317-325.

[9] Park C.W. and Kwon W.H., 2004, Simple and robust sensorless vector control of induction motor using stator current based MRAC, *Electric Power Systems Research.*, **71** 257-266.

[10] Peng, F.Z. and Fukao, T.,1994, Robust Speed Identification for Speed-Sensorless Vector Control of Induction Motors. *IEEE Trans. Indus. Applica.*. **30**, 1234-1240.

[11] Sabanovic, A. and Izosimov, D.B., 1981, Application of Sliding Modes to Induction Motor Control, *IEEE Trans. Indus. Applica.*, **IA-17**, 41-49.

[12] Schauder C., 1992, Adaptive Speed Identification for Vector Control of Induction Motors without Rotational Transducers, *IEEE Trans. Indus. Applica.*, **28**, 1054-1061.

[13] Slotine, J.J.E. and Li, W. (1991). *Applied nonlinear control*. Prentice-Hall, New Jersey.

[14] Vas, P., 1994, *Vector Control of AC Machines*. Oxford Science Publications, Oxford.

INDEX

N

O

P